MW01703380

Mass Spectrometric Analysis of Oligonucleotide and DNA Adducts

Qing Liao

Mass Spectrometric Analysis of Oligonucleotide and DNA Adducts

Computational, Enzymatic and Mass Spectrometric Methods for DNA Adducts Analysis in Cancer Related Research - The Past, Present and Future

VDM Verlag Dr. Müller

Impressum/Imprint (nur für Deutschland/ only for Germany)
Bibliografische Information der Deutschen Nationalbibliothek: Die Deutsche Nationalbibliothek verzeichnet diese Publikation in der Deutschen Nationalbibliografie; detaillierte bibliografische Daten sind im Internet über http://dnb.d-nb.de abrufbar.
Alle in diesem Buch genannten Marken und Produktnamen unterliegen warenzeichen-, marken- oder patentrechtlichem Schutz bzw. sind Warenzeichen oder eingetragene Warenzeichen der jeweiligen Inhaber. Die Wiedergabe von Marken, Produktnamen, Gebrauchsnamen, Handelsnamen, Warenbezeichnungen u.s.w. in diesem Werk berechtigt auch ohne besondere Kennzeichnung nicht zu der Annahme, dass solche Namen im Sinne der Warenzeichen- und Markenschutzgesetzgebung als frei zu betrachten wären und daher von jedermann benutzt werden dürften.

Coverbild: www.purestockx.com

Verlag: VDM Verlag Dr. Müller Aktiengesellschaft & Co. KG
Dudweiler Landstr. 99, 66123 Saarbrücken, Deutschland
Telefon +49 681 9100-698, Telefax +49 681 9100-988, Email: info@vdm-verlag.de

Herstellung in Deutschland:
Schaltungsdienst Lange o.H.G., Berlin
Books on Demand GmbH, Norderstedt
Reha GmbH, Saarbrücken
Amazon Distribution GmbH, Leipzig
ISBN: 978-3-639-14685-1

Imprint (only for USA, GB)
Bibliographic information published by the Deutsche Nationalbibliothek: The Deutsche Nationalbibliothek lists this publication in the Deutsche Nationalbibliografie; detailed bibliographic data are available in the Internet at http://dnb.d-nb.de.
Any brand names and product names mentioned in this book are subject to trademark, brand or patent protection and are trademarks or registered trademarks of their respective holders. The use of brand names, product names, common names, trade names, product descriptions etc. even without a particular marking in this works is in no way to be construed to mean that such names may be regarded as unrestricted in respect of trademark and brand protection legislation and could thus be used by anyone.

Cover image: www.purestockx.com

Publisher:
VDM Verlag Dr. Müller Aktiengesellschaft & Co. KG
Dudweiler Landstr. 99, 66123 Saarbrücken, Germany
Phone +49 681 9100-698, Fax +49 681 9100-988, Email: info@vdm-verlag.de

Copyright © 2009 by the author and VDM Verlag Dr. Müller Aktiengesellschaft & Co. KG and licensors
All rights reserved. Saarbrücken 2009

Printed in the U.S.A.
Printed in the U.K. by (see last page)
ISBN: 978-3-639-14685-1

ACKNOWLEDGEMENTS

The research in this book was conducted under the direction of Professor Paul Vouros of Northeastern University in Boston, Massachusetts, who is instrumental in my intellectual growth. He has taught me how to think deeply, to be methodological and to strive for perfection. He enabled me to plant my feet firmly in research and fostered my strong interest in DNA adduct study and carcinogenesis. I have learnt from him thoughtful approaches to experimental work. He helped me in continuously formulating, adjusting, questioning, readjusting and revising the manuscript for this book. He has given me invaluable scientific insights and inspiration. Without his sound advice and guidance, this book would not have been possible.

Dr. Chang Shen is the co-developer of the GenoMass program. Dr. Shen obtained his Ph.D. in Electrical Engineering from the Swiss Federal Institute of Technology (EPFL) in Lausanne, Switzerland and and is currently a principle engineer in US Genomics at Woburn, Massachusetts. His interest in life science has led him to devote significant amount of time and efforts to the project. Dr. Shen is largely responsible for the software/data handling development of this research work.

I would like to thank some present and past members in the Vuoros group of Northeastern University, Boston, for their help and constructive comments, in particular, James Glick, Susan Schiavo, Dayana Argoti, Caroline Ceailles, and Christine Andrews. Professor Norman Chiu at University of North Carolina in Greensboro offered his valuable feedback and comments. I also apprectiate Professor Patricia Mabrouk, Professor Geoffrey Davies, and Professor David Forsyth of Northeastern University for carefully reading my manuscript and providing very helpful suggestions regarding the work.

One of my former associates at Harvard University, Dr. Yi Chen, originally from Tsinghua University in China, deserves special credit for contributing to a portion of the research.

Many thanks to Professor Yan Liao, a librarian at University of Wisconsin-Stevens Point, who has helped me with editing the manuscript and the references with great attention to detail.

Funding for the research in this work was provided by the National Institute of Health (NIH), under grant #R01CA69390.

To My Dearest Family:

My daughter Vivian
My husband Chang Shen
My mom and dad
My sister Yan
And my 98 year old grandma

献给我至爱的家人

女儿薇薇
先生沈畅
爸爸妈妈
妹妹
和 98 岁的外婆

TABLE OF CONTENTS

LIST OF FIGURES

LIST OF TABLES

Page

LIST OF ABBREVIATIONS

μESI-MS	Micro-electrospray mass spectrometry
μg	Microgram
μL	Microliter
μM	Micromolar
2,4DMA	2,4-Dimethylaniline
2CA	2-Chloroaniline
2-NT	2-Nitrotoluene
3'PDE	Snake venom phosphodiesterase
4,5-EDE	4,5-Epoxy-2(E)-decenal
4-HNE	4-Hydroxy-2-nonenal
4MA	4-Methylaniline
4-OHE	4-Oxo-2-hexanenal
4-OHEN	4-Hydroxy-equilenin
4-ONE	4-Oxy-2-nonenal
5-HMU	5-Hydroxymethyluracil
5-nitro-BNT	5-Nitrobenzo[b]naphtha[2,1-d]thiophene
5'PDE	Bovine spleen phosphodiesterase
7-HEG	7-(2-Hydroxyethyl)guanine
8-OH-dG	8-Hydroxy-2'-deoxyguanosine
8-oxodG	7,8-Dihydro-8-oxo-2'-deoxyguanisine
A	Adenine
AA	Aromatic amines
AAAF	N-acetoxy-2-acetylaminofluorene
AAF	Acetylaminofluorene
ABP	4-Aminobiphenyl
ACN	Acetonitrile
ADHD	Attention deficit hyperactivity disorder
AFB	Aflatoxin B
AGCT	Melphalan alkylated tetra-mer
AMS	Accelerated Mass Spectrometry
APCI	Atmospheric pressure chemical ionization
ATP	Adenosine triphosphate
BaA	Benzo[a]anthracene
BaP	Benzo[a]pyrene
BD	1, 3-Butadiene
BDAB	Butyldimethylammonium bicarbonate
BDO	1,2-Epoxybutene
BkF	Benzo[b]fluoranthene
BLAST	Basic Local Alignment Search Tool
BLM	Bleomycin
BPADGE	Bisphenol A diglicydyl ether
BPDE	Benzo[a]pyrene diol epoxide
BZ	Benzidine
C	Cytosine
C8-AAF-dGuo	N-acetyl-N-(guan-8-yl)-2-aminofluorene
capLC-μMS	Capillary LC-microspray MS
CDC	Center for disease control and prevention
CE	Capillary electrophoresis

CEC	Capillary electrochromatography
CE-MS	Capillary electrophoresis – Mass Spectrometry
CI	Chemical ionization
CID	Collision induced dissociation
CNL	Constant neutral loss
COMPAS	Comparative sequencing algorithm
CR	Crotonaldehyde
ctDNA	Calf thymus DNA
CZE	Capillary zone electrophoresis
Da	Dalton
dA	Deoxyadenosine
dC	Deoxycytosine
DCB	Dichlorobenzidine
DEB	Depoxybutane
dG	Deoxyguanosine
DHP	6,7-Dihydro-7-hydroxy-1-hydroxymethyl-5H-pyrrolizine
DI	Deionized
DMBA	7,12-Dimethylbenz[a]anthracene
DMF	Dimethylformamide
dN	Deoxynucleoside (dA, dG, dC or dT)
DNA	Deoxyribonucleic acid
DNase I	Deoxyribonuclease I
dT	Deoxythymidine
EH	2,3-epoxy-4-hydroxynonenal
EI	Electron impact ionization
EIC	Extracted ion chromatogram
ELISA	Enzyme-linked immunosorbent assays
ENU	N-ethyl-N-nitrosourea
epsilon	N6-ethenoadenine
ES	Electrospray
ESI	Electrospray ionization
ESI-ITMS	Electrospray ionization ion trap mass spectrometry
ESI-LC/MS	Electrospray ionization liquid chromatography mass spectrometry
eV	Electron volts
FS	Fitness value
FTICR	Fourier transform ion cyclotron resonance
G	Guanine
GC	Gas chromatography
GC/NICI/MS	Gas chromatography/negative ion/chemical ionization mass spectrometry
GC-EC-NICI HRMS	Gas chromatography-electron capture negative ion chemical ionization-high resolution-mass spectrometry
GC-EC-NICI-MS	Gas chromatography-electron capture negative ion chemical ionization- mass spectrometry
GC-MS	Gas chromatography - mass spectrometry
GSCH2OAc	S-(1-Acetoxymethyl)glutathione
HAA	Heterocyclic aromatic amines
HEG	(2-Hydroxyethyl)guanine
HFIP	1,1,1,3,3,3-Hexafluoroisopropanol

HOCl	Hypochlorous acid
HOEtVal	Hydroxyethlvaline
HPLC	High-performance liquid chromatography
HPLC-MS	High-performance liquid chromatography - mass spectrometry
HPNE	4-Hydroperoxy-2-nonenal
HRMS	High-res MS
HX	Hexenal
IARC	International Agency for Research on Cancer
ID	Inner diameter
IP-RP-HPLC	Ion-pairing reversed phase HPLC
IQ	2-Amino-r-methylimidazo[4,5-f]quinoline
IR	Infrared spectroscopy
kV	Kilovolt
LC	Liquid chromatography
LC-MS	Liquid chromatography - mass spectrometry
LOD	Limits of detection
LTA4	Leukotiene A4
m/z	Mass-to-charge ratio
M1G	Pyrimido[1,2-a]-purin-10(3H)-one
MALDI	Matrix assisted laser desorption ionization
MDA	Malondialdehyde
MeIQx	2-Amino-3,8-dimethylimdazo[4,5-f]quinxaline
MMS	Methyl methanesulfonate
MNU	N-methyl-N-nitrosourea
Mr	Relative molecular mass
MRM	Multiple reaction monitoring
MS	Mass spectrometry
MS/MS	Tandem mass spectrometry (two stages)
MS^n	Tandem mass spectrometry (n stages)
MW	Molecular weights
nanoLC-μESI-MS	Nano liquid chromatography micro electrospray mass spectrometry
NC	Nitrochrysene
NCBI	National Center for Biotechnology Information
NDEA	N-nitrosodiethanolamine
NDELA	N-nitrosodiethanolamine
NF	Nitrofluoranthene
NIH	National Institutes of Health
NICI	Negative ion chemical ionization
NL	Neutral loss
NMR	Nuclear magnetic resonance spectroscopy
NNAL	4-(Methylnitrosamino)-1-(3-pyridyl)-1-butanal
NNK	4-(Methylnitrosamino)-1-(3-pyridyl)-1-butanone
NNN	N'-nitrosonornicotine
NP	Nitropyrene
OTA	Ochratoxin A
PAH	Polycyclic aromatic hydrocarbon
PC	Polycarbonate
PCB	Polychlorinated biphenyls

PCR	Polymerase chain reaction
PGE	Phenyl glycidyl ether
PhIP	2-Amino-1-methyl-6-phenylimidazo[4,5-b]pyridine
POB	Pyridyloxobutyl
ppm	Parts per million
PSDVB	Polystyrene divinylbenzene
PVP	Polyvinylpyrrolidone
RNA	Ribonucleic acid
ROS	Reactive oxidizing species
RP	Reverse phase
RSD	Relative standard deviation
RT	Retention time
S/N	Signal to noise ratio
SFS	Synchronous fluorescence spectroscopy
SIM	Single ion monitoring
SO	Styrene oxide
SOMA	Short oligonucleotide mass analysis
SOS	Simple oligonucleotide sequencer
SPE	Solid phase extraction
SRM	Selected reaction monitoring
T	Thymine
TAM	Tamoxifen
TEA	Triethylammonium acetate
TEAA	Triethyammonium acetate
TEAB	Triethylammonium
TIC	Total ion current
TLC	Thin-layer chromatography
ToF	Time-of-flight
UV	Ultraviolet
VC	Vinyl chloride
WBC	White blood cells

Chapter 1

Advances in Analysis of DNA Adducts by Separation Techniques and Mass Spectrometry in Cancer-Related Research: A Perspective

1.1 INTRODUCTION

1.1.1 Initiation of Cancer

Cancer is a genetic disease that affects millions of people and is, at least in part, due to DNA damage caused by chemical, viral or physical carcinogens. Among all diseases, cancer is the second lethal killer after cardiovascular disease. It kills nearly 1,500 Americans a day. In year 2004 alone, the deaths caused by cancer were 553,888 (NIH and American Cancer Society). The estimate of cancer deaths for 2007 by the American Cancer Society are 559,650 despite the fact the federal government alone has spent $69 billion on cancer research since 1971. The most common types of cancer are breast cancer, prostate cancer, and colorectal cancer. Cancer can be attributed to a single underlying malady of the genetic program that directs the lives of our cells. When organisms fall prey to uncontrolled proliferation of defective, undifferential cells, cancer occurs. A so-called carcinogenic event is usually set off by some external factors such as exposure to sun-light, x-rays, smoking, and dangerous chemicals. Exposure to genotoxic compounds can alter DNA sequence, lead to mutated genes and give rise to genetic instability. The myriads of alterations within the cell include switching on or off genes, aberrant protein expression or altering cell cycle control.

There is overwhelming evidence that mutations in relevant target sequences, such as oncogenes or tumor suppressor genes, are associated with the carcinogenic process. A number of cancer-susceptible genes have been identified including *p53*, *ras*, *rb*, *BRCA1*, *BRCA2*, *hprt*, *aprt* and *dhfr*. The mechanism of mutation is quite complex, it can be frameshift, base deletion or base substitution that occurs as a result of chemical reactions between genotoxic compounds and DNA. It has been estimated that every gene in our DNA is damaged some 10 billion times in our lifetime yet the rate of the mutations is far lower. This is mainly due to the ability of the cell to repair its damaged DNA. However, if not repaired, mutation can occur. Therefore, it is of paramount importance to a cell to maintain the integrity of DNA structure to have less chance of mutation and likelihood of cancer occurrence.

1.1.2 Biomarkers

Keys to the success in the fight against cancer include prevention, early detection and more effective treatment. In recent years, there were declines in annual cancer deaths owing largely to screening efforts. We are seeing some real dividends as earlier detection was made possible by advances in technology. What are called biomarkers have played a significant role in this.

Biomarkers are distinctive biological indicators of molecular and cellular information that can mark the presence or progression of a disease. They can be acquired from patients from host cells/tissues of tumors and body fluids. Detecting biomarkers specific to a disease can help in identification, diagnosis, and treatment of those with the disease as well as those who might be at risk but are as yet asymptomatic. In cancer research specifically, the study of cancer biomarkers may help to: i) identify who is at risk of the disease, ii) diagnose the disease at an early stage, iii) select the best treatment, and iv) monitor response to treatment. To this date, cancer biomarkers have included indicators such as: carcinogen-protein adducts, carcinogen-DNA adducts, chromosomal aberrations, polymorphisms in drug metabolizing enzymes, and host DNA repair capacity.

1.1.3 DNA Adducts as Biomarkers in Cancer Research

A major goal of DNA adduct studies is to use the information in predicting human cancer risk. In our daily life we are constantly exposed to genotoxic compounds that may interact with cellular DNA to form chemical products called DNA adducts, where adducting species (carcinogens) are covalently bonded to DNA. DNA adducts formation is generally considered one of the key events in tumor initiation during chemical carcinogenesis. Considerable epidemiological and experimental evidence has indicated that exposure to most carcinogens results in damage to the structural integrity of DNA. DNA adducts are the precursor lesions for mutation, as mutations often occur at (or close to) the carcinogen adduction sites (Singh et al., 2006; Farmer et al., 2005; Turesky and Vouros, 2004). DNA adduct formation often leads to malfunction of DNA that includes major distortions in DNA, and even formation of crosslinks between two strands. The majority of DNA adducts are eliminated by DNA repair processes but, if not repaired, may result in misreplication, inhibition of DNA synthesis, or termination of transcription that will lead to mutation in important growth-controlling genes or loci, resulting in mutagenesis. Alterations in DNA sequence may also occur when adducts are subjected to erroneous repair, as damage to DNA can increase the frequency of miscopying of DNA, which may lead to apoptosis or frameshift mutation. The consequence of mutagenesis is aberrant cellular growth and ultimately cancer. Since formation of DNA adducts is a potential first step in the initiation of cancer, their measurement can provide at the very least an estimate of internal exposure and ideally a measure of biological outcome. For these reasons, DNA adducts are viewed as biomarkers to assess genetic damage which may also be utilized in the molecular epidemiology of cancer for risk assessment.

Evidence of the mutagenicity of DNA adducts came from studies involving incorporation of single adducts into defined DNA sequences using site-specific techniques, as well as analyzing induced mutational spectra in specific genes in cells of treated animals. Mutational spectra are used to provide information on the frequency and types of mutations that arise from a particular DNA-damaging compound. Numerous studies support the hypothesis that DNA adduct levels in target and in surrogate tissues are appropriate biomarkers of DNA damage induced by carcinogen exposure. Extensive evidence gained through carcinogenicity testing in experimental animals suggests that while DNA adduct formation alone is insufficient to cause tumor formation, it is at least one prerequisite (Garner, 1998; Hemminki, 1993; Hemminki et al., 2000; Santella et al., 2005; Vulimiri and DiGiovanni, 2002).

Although the data are less definitive in humans, it appeared that the relationship between DNA adduct levels and tumor incidences following lifetime administration of different doses of chemical carcinogens to experimental animals and the mechanisms derived from such studies can be extrapolated to provide the best basis for human cancer risk assessment.

The quantitative correlation between DNA adducts formation and cancer risk is still unclear. There are uncertainties concerning the biological significance of low levels of DNA adduct formation in particular. The scientific community has yet to agree on whether there is a threshold of DNA adducts below which there is no measurable biological effect. It is unlikely that a single 'universal' value could be given to such an acceptable level of adducts, as their mutational effectiveness varies according to the nature of the carcinogen and the chemical structure of the DNA adduct. There is no simple algorithm for translating DNA adduct levels into cancer risk in a particular tissue. It is generally accepted by many regulatory authorities that the dose-response relationship for genotoxic carcinogens does not have a threshold and thus the presence of any amount of exposure presents a carcinogenic risk. The dose-response relationship for genotoxic carcinogens is likely non-linear as it is not always possible to detect any increase in mutations above background at administered doses of genotoxic agents which produce detectable adducts. On the other hand, the lack of adduct formation might suggest that no mutation could occur, and demonstration of the inability of a chemical to produce adducts may be used to eliminate the possibility that the compound is a genotoxic carcinogen. This may have a major influence on regulatory decisions on such compounds. Although the relative roles of chemical reactivity in the formation of a lesion and the carcinogenic potency of a particular lesion in the establishment of clonal growth advantage remain unclear, nonetheless as a means to cancer prevention it is crucial to decrease the amount of DNA damage. Put it in simple terms, any modification to genetic material cannot be overlooked.

The advantage of using DNA adducts as biomarkers and comparison with environmental monitoring data is that it can provide several aspects of assessment. These include the information leading to exposure levels, absorption, distribution, metabolic activation (or inactivation), genetic susceptibility and DNA repair capacity. There is, however, consensus that the use of adduct data in risk extrapolation has the greatest value when adduct structure has been characterized and the roles of adduct removal and biological relevance of specific adducts are understood. The biological potential of a given DNA adduct depends on its mutagenicity, ability to be repaired, location within a target gene, and the nature of the target gene. Total DNA adduct levels, and in some cases specific adducts, have been correlated with in vitro mutation, chromosomal aberrations and generally with carcinogenicity. Thus, quantitation of DNA adducts in human cells provides a useful parameter of risk assessment in terms of monitoring human carcinogen exposure, the determination of a biologically effective dose, and individual cell type-specific DNA repair capacity. Detection and quantitation of DNA adducts in human tissues also confirms epidemiological associations of cancer and risk factors and provides information on the identification of carcinogenic hazard and quantitative risk assessments of accumulative genetic damage.

Risk assessment with DNA adducts as biomarkers does have limitations. For example it does not take into account inter-individual variation in absorption, metabolism, excretion and bioavailability of the carcinogens. The presence of relatively high levels of certain endogenous DNA adducts, such as oxidized bases induced by endogenously formed oxygen radicals, may further complicate the interpretation of data. This makes further characterization necessary, including structure-activity relationship studies, mechanistic studies on species differences in metabolism, and investigation of genetic polymorphisms.

Although there is a multitude of factors affecting the interpretation of the relationship between DNA adducts levels and carcinogenesis, it should be realized that when DNA adducts, including decomposed forms of bases, are detected with remarkable frequency, the following situations have arisen: i) there are mutagenic chemicals, sources of dangerous electromagnetic waves or serious stresses to cause DNA mutation around our lives, ii) the possibility of falling victim to serious diseases such as cancers caused by changes in the DNA sequences is increased drastically. We are required to decrease the quantity of mutagenic subjects in our surroundings and, if possible, to remove them completely. On the other hand, we should investigate the mechanisms expressing toxicity for each DNA adduct and also reveal the strength of the toxicity.

In summary, since the formation of specific DNA modifications appears to be a critical event in carcinogenesis, measurement of carcinogen-DNA adducts should provide biologically relevant information on the net result of exposure, absorption, metabolism, DNA adduct formation and DNA repair. Thus, analysis of DNA adduct levels may be one of the best tools available to characterize exposure to complex mixtures of genotoxic chemicals in different environmental and occupational exposure settings. Consequently, the study of DNA adducts as biomarkers for cancer involves determination of the nature and extent of adduct formation, and if possible also the position of these adducts within the DNA structure. It is anticipated that the characterization of DNA adducts in conjunction with other biological information, e.g., gene expression analysis, mutation analysis, etc., will play a greater role in the assessment of carcinogenic hazard and possibly of risk in the foreseeable future.

1.1.4 Classes of Genotoxic Compounds

There are two types of DNA-damaging agents, the exogenous compounds and the endogenous compounds. The exogenous compounds are what are commonly known as carcinogens that have been demonstrated by epidemiological studies to be genotoxic chemicals of a wide range. They are chemicals that can be divided into two groups: the genotoxic, which cause damage to DNA, and the non-genotoxic, which cause no direct damage but indirectly initiate DNA damage because of their metabolic pathway. They often require activation to reactive metabolites that bind to nucleophilic centers in DNA to form covalent adducts. The exogenous compounds include mainly these classes of compounds: polycyclic aromatic hydrocarbons (PAHs), nitro-PAHs, N-nitrosamines, aromatic amines (AA), heterocyclic aromatic amines (HAA), and aflatoxins. DNA adducts can also be described in terms of their hydrophobicity and divided into small or polar adducts and bulky or non-polar adducts.

The sources of these carcinogens can be dietary, life style related (such as cigarette smoking), medicinal, occupational and environmental/industrial (air, water) and UV radiation. For example, carcinogens found in the diet include mycotoxins and N-nitroso compounds. There are also compounds generated by the cooking process including HAA, PAHs and acrylamide. Many potential human genotoxic carcinogens are formed following the heating of carbohydrate rich foods to high temperatures. Environmental air pollution is another major source of carcinogens. In highly industrialized cities, elevated levels of PAHs are found. Lifestyle is another important factor when considering the total scheme of exposure events, of which cigarette smoke is the major culprit. In summary, one may encounter multiple sources of exposures in daily life. Cigarette smoking, environmental tobacco smoke, exhaust by-products, foods, etc. may all contribute to a person's cumulative risk of cancer.

For example, cigarette smoke contains ~ 4000 chemicals, about fifty compounds of which have been identified as either animal or human carcinogens according to the International Agency for Research on Cancer (IARC). The major players are PAHs and the alkylating tobacco-specific nitrosamines which most likely lead to lung cancer. In addition, cigarette smoke can generate reactive oxidizing species (ROS) or free radicals that can hydroxylate DNA bases. They can induce DNA single strand breaks in vitro as well as cause oxidative DNA damage in cultured human cells. Exogenous modifications often involve aromatic or polycyclic aromatic metabolites to form bulky DNA adducts at concentrations ranging from 1 adduct in 10^6 to 10^9 unmodified nucleotides.

Endogenous compounds are derived from endogenous electrophiles released from normal cellular function such as metabolism of nutrients and other natural dietary components. Many of these agents are highly reactive, and thus do not need further metabolic activation. Various types of endogenous DNA damage include those from DNA instability, errors in replication and repair, oxidatively damaged bases and adducts derived from reaction of bases with aldehydic lipid peroxidation products. Endogenous DNA damage is present at relatively high levels in mammalian tissues and, although it is well documented that it can be carcinogenic. The investigations have just commenced into how to assess the damage done by them in terms of human carcinogenesis. Endogenous DNA adducts are formed by small modifications, for example by oxygen radicals, and are normally found at concentrations of 1 adduct in 10^4 to 10^7 unmodified nucleotides.

More than 800 different compounds have been evaluated by IARC. Over 75 specific adducts or closely related groups of adducts have been determined to be human carcinogens (Hemminki, 1993; Hemminki et al., 2000). Many compounds form the same adducts and also diastereomers or enantiomers of the same structural type. Hemminki and Garner each compiled a table of DNA adducts detected in human tissues and the method of detection (Garner, 1998; Hemminki et al., 2000). The sources from which the DNA adducts can be

e usually total white blood cells (WBC) or peripheral blood lymphocytes, and to a
t skin, placenta, kidney, liver, lung, breast and pancreas.

Table 1.1. Genotoxic Compounds Detected in Human Tissues and Methods of Detection (adapted and modified from Garner, 1998 and Hemminki et al., 2000).

Chemical	Tissue	Method
4-Aminobiphenyl	Bladder, lung	Postlabeling, GC/MS
4,4'-Methyleneebis(2-chloroaniline)	Bladder	Postlabeling
Methylating agents	Lymphocytes, liver, lung, bronchus, WBC	IAC/postlabeling, LC-MS
Aflatoxin B_1	Urine, lung, kidney, liver	Immunoassay, Fluorescence, LC-MS
Cisplatin	Lymphocytes, WBC	HPLC/immunoassay, LC-MS
Benzo(a)pyrene	WBC, lung	Fluorescence, LC-MS
Cigarette smoke	Lung, bladder, breast lymphocytes, cervix, placenta, buccal mucosa	Postlabeling
Foundry fumes	Lymphocytes	Postlabeling
Air pollution	Lymphocytes	Postlabeling
MeIQx	Colon, kidney, rectum	Postlabeling, LC-MS
PhIP	Colon	GC/MS, LC-MS
Aristolochic acid	Kidney	Postlabeling
Malondiadehyde	Lymphocytes, liver, pancreas, breast, WBC	GC/MS, Postlabeling
Styrene oxide	Lymphocytes, WBC	Postlabeling, LC-MS
4-Hydroxynonenal	Lymphocytes, liver	Immunoaffinity/ Postlabeling
Etheno A, C, G (vinyl chloride)	Liver, WBC, pancreas	IA/Postlabeling, LC-MS
Acrolein	Liver, WBC, mammary gland	Postlabeling, LC-MS
Crotonaldehyde	Liver, WBC, mammary gland	Postlabeling, LC-MS
Tamoxifen	WBC, endometrium	Postlabeling, LC-MS
8-OxodG	WBC, pancreas	EC, LC-MS
Benzidine	WBC	Postlabeling, LC-MS
Butadiene	Lymphocytes	Postlabeling, LC-MS
UV photoproducts	Skin	Postlabeling

General classes of these carcinogens include alkylating agents, aromatic amines, epoxides, PAHs, nitro-PAHs, food-derived HAAs, aldehydes, allylic compounds, and chemotherapeutic drugs. Under the PAHs category the most studied is benzo[a]pyrene (BaP) and its metabolite benzo[a]pyrene diol epoxide (BPDE), to a lesser extent benzo[a]anthracene (BaA) and benzo[b]fluoranthrene (BkF) (Beland et al., 2005; Chen et al., 2005; Churchwell et al., 2006b; Marzilli et al., 2001; Singh et al., 2006; Wang et al., 2003). In the class of nitro-PAHs there are 4-nitropyrene (4-NP), 6-nitrochrysene (4-NC) and 3-nitrofluoranthrene (3-NF). In the life style related carcinogens the methylating nitrosamines are mostly found in tobacco smoke. Some of the most extensively studied are 4-(methylnitrosamino)-1-(3-pyridyl)-1-butanone (NNK) and its metabolite 4-

(methylnitrosamino)-1-(3-pyridyl)-1-butanal (NNAL) and N'-nitrosonornicotine (NNN) (Guza et al., 2006; Haglund et al., 2002; Hecht et al., 2004; Lao et al., 2005, 2006a, 2006b; Rajesh et al., 2005; Tretyakova, 2003, Upadhyaya et al., 2003; Villata et al., 2006). Under the category of PAAs there are 2-aminofluorene, benzidine (BZ) and 3',3'-dichlorobenzidine (DCB). The occupational environmental carcinogens often studied are arylamines (AA), nitroarenes, polychlorinated biphenyls (PCB) and compounds such as urethane and N, N-dimethylformamide (DMF). In the class of arylamines the most studied are 4-aminobiphenyl (4-ABP) and its analogs such as 2ABP and 3ABP. Less studied are 2-chloroaniline (2CA) and 4CA, 4-methylaniline (4MA), 2,4-dimethylaniline (2,4DMA) and 2,6DMA. In the class of nitroarenes there are 2-chloronitrobenzene and 4-chloronitrobenzene, 2-nitrotoluene (2-NT), 4-NT and nitrofluorenes. There are thioarenes such as 5-nitrobenzo[b]naphtha[2,1-d]thiophene (5-nitro-BNT), dihaloalkanes such as 1,2-dihaloethane. In diet there are carcinogens such as mycotoxin A, N-nitrosodiethanolamine (NDELA) and furan. Some cancer chemotherapeutic agents have a secondary carcinogenic effect. One class most often studied is nitrogen mustards which are bifunctional alkylating agents. Two nitrogen mustards, melphalan and tamoxifen (TAM), are model carcinogens in many studies. In addition, there are some sugar adducts and antibiotics which have some carcinogenic effects. One radiation related carcinogen is 5-hydroxymethyluracil (5-HMU).

There have been many studies on model carcinogens including platinum-DNA adducts such as cisplatin (McDonald et al., 2005; Qu et al., 2003, 2004; Tu et al., 2004; Warnke et al., 2004), estrogens, including the metabolite 4-hydroxy-equilenin (4-OHEN), 4-hydroxy-estradiol and 4-hydroxy-estrone (Borges et al., 2004; Chen et al., 2000; Debrauwer et al., 2003; Embrechts et al., 2003; Li et al., 2004) and styrene oxide (SO) (Edler et al., 2005; Shao et al., 2002, Tarun and Rusling, 2005a, 2005b; Tarun et al., 2006; Yang et al., 2005). Other important carcinogens receiving attention are 2-amino-1-methyl-6-phenylimidazo[4,5-b]pyridine (PhIP), 2-amino-r-methylimidazo[4,5-f]quinoline (IQ), aflatoxin B (AFB), acrylamide, N-methyl-N-nitrosourea (MNU), 2-amino-3,8-dimethylimdazo[4,5-f]quinxaline (MeIQx), phenyl glycidyl ether (PGE), bisphenol A digylcocidy ether, choloroambucil, ochratoxin A (OTA), hydroxyethylvaline (HOEtVal), 7-(2-hydroxyethyl)guanine (7-HEG), S-(1-Acetoxymethyl)glutathione (GSCH2OAc), bleomycin (BLM), hypochlorous acid (HOCl), 1,2-epoxybutene (BDO), redelline, and leukotiene A4 (LTA4). Aacetaldehyde (AA), crotonaldehyde (CR), formaldehyde, acrolein and 1, 3-butadiene (BD), depoxybutane (DEB) and methyl methanesulfonate (MMS) are also among the carcinogens that appear often in the literature.

The most extensively studied class of carcinogens perhaps are lipid hydroperoxidation products and ROS generated from them (Churchwell et al., 2002, 2006a, 2006b; Doerge et al., 1998, 2000; Jian et al., 2004; Xu and Blair, 2004). Among these are etheno DNA adducts such as N6-ethenoadenine (epsilon A) and vepsilonA, MDA, vinyl halides such as vinyl chloride (VC), pyrimido[1,2-a]-purin-10(3H)-one (M1G) and to a lesser extent, 4-oxy-2-nonenal (4-ONE), 4,5-epoxy-2(E)-decenal (4,5-EDE), 2,3-epoxy-4-hydroxynonenal (EH), 4-hydroxy-2-nonenal (4-HNE), 4-oxo-2-hexanenal (4-OHE) and, 4-hydroperoxy-2-nonenal (HPNE) etc. (Gonzalez-Reche et al., 2002; Ham et al., 2004; Hillestrom et al., 2004, 2005, 2006a, 2006b, 2006c; Kaddiska et al., 2005; Lee and Blair, 2004; Wu et al., 1999). Chen and Chang (2004), Chen and Chiu (2005), Chen et al. (2003, 2004a, 2004b) and Chen et al. (2005) of the National Chung Cheng University, Taiwan conducted an extensive study of lipid peroxide products, particularly etheno DNA adducts, using gas chromatography/negative ion/chemical ionization mass spectrometry (GC/NICI/MS). In more recent studies, they explored the use of LC-MS and CE-MS (Chen and Chang, 2004; Chen and Chiu, 2005). They also reviewed the formation, analysis and repair of exocyclic etheno DNA adducts (Chen, 2004).

Among the various forms of oxidative lesions, the ROS 7,8-dihydro-8-oxo-2'-deoxyguanisine (8-oxodG) is the oxidation product from above species. 8-oxodG is the most extensively studied biomarker for oxidative damage (Hofer et al., 2005; Li et al., 2005; Matter et al., 2006; Rajesh et al., 2003; Shiota et al., 2002; Wang et al., 2005; Xi et al., 2004) often detected in urine and has been shown to be carcinogenic. Urinary excretion of 8-oxodG has been used extensively as a noninvasive biomarker of oxidative DNA damage in humans.

1.1.5 Mechanism of DNA Adduct Formation

Carcinogens are usually electrophilic species or become electrophilic after metabolism in vivo, and bind covalently to the nucleophilic sites of DNA. In general, adduction can occur on the DNA base as well as on the phosphate and deoxyribose moieties, to the nucleophilic sites in the nucleic acid structure. On the bases, the nitrogen and oxygen atoms are active sites. Of the four DNA bases, the one mostly susceptible to carcinogen modification is guanine (G) and the C8 on the guanine is the place attacked most often by carcinogens. The N7 position of G is predominantly modified by alkylation (methylation or ethylation) agents, while aromatic amines and PAHs prefer the C8 and N2 position, respectively.

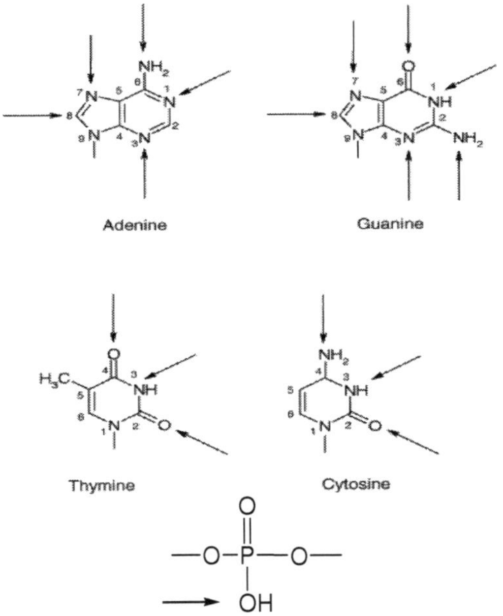

Phosphate

Figure 1.1. Potential sites of addition reactions for the DNA bases (adapted and modified from Esaka et al., 2003).

Table 1.2. Some common carcinogens and their binding sites on DNA (adapted and modified from Singh et al., 2006).

Adduct Bonding Site	Genotoxic Compounds Bound to DNA
N-1, N^2	Lipid peroxidation products Malondialdehyde (MDA) 4-Hydroxy-2-nonenal Crotonaldehyde (CR) 2-Hexenal (HX) Vinyl chloride (VC) Ethanol Estrogen *N*-Nitrosodiethanolamine
N^2	2-Amino-r-methylimidazo[4,5-f]quinoline (IQ) 2-Amino-1-methyl-6-phenylimidazo[4,5-b]pyridine (PhIP) 2-Amino-3,8-dimethylimdazo[4,5-f]quinxaline (MeIQx) Tamoxifen (TAM) Ethanol Phenyl glycidyl ether (PGE) Benzo[a]pyrene (BaP) 2-Aminofluorene 2-Acetoxy-Acetyl-Aminofluorene (AAAF)
N^2, N-3	Lipid peroxidation products 4-Hydroxy-2-nonenal
O^- (on phosphate)	Ethylating agents PGE
C-8	Hydroxyl radicals Ethanol 4-Aminobiphenyl (ABP) PhIP MeIQx IQ Peroxynitrite 2-Aminofluorene AAAF
N-7	Methylating agents Ethylating agents Ethylene 1, 3-Butadiene (BD) Acrylamide BaP Benzo[a]pyrene diol epoxide (BPDE) 7,12-Dimethylbenz[a]anthracene (DMBA) Styrene oxide (SO) PGE Aflatoxin B (AFB) Estrone
O^6	N-Nitrosodiethanolamine (NDEA) N-Methyl-N-nitrosourea (MNU) N-Ethyl-N-nitrosourea (ENU) 4-(Methylnitrosamino)-1-(3-pyridyl)-1-butanone

1.2 STRATEGIES FOR THE ANALYSIS OF DNA ADDUCTS

1.2.1 Analytical Challenges

The major challenge in analyzing DNA adducts lies in the extremely high sensitivity and selectivity that is required to detect and quantify these compounds in vivo where they are present in only minute amounts. The damaged bases in DNA of living organisms are usually in the range of 1 in 10^6 to 10^8 and can be as low as 1 in 10^{11} normal nucleotides. This means that one may expect ca. 10 pg of adduct in 1 mg of DNA, effectively searching for a needle in a haystack. Assuming an analytical mass detection limit of 10 pg and setting as a goal the detection of 1 adduct in 10^8 nucleotides, this would translate into a requirement of ca. 1 g of tissue sample (considering that 1 g of tissue may contain 1 mg of DNA). In practice, it is desirable to have about a 10-fold excess of analyte. Thus there appears to be two alternatives: either have a sample 10 g of tissue or improve the detectability by a factor of 10 or more. The availability of DNA samples from human is actually quite restricted. DNA human samples mainly are obtained from white blood cells and the yield is low – typically 1 mL of blood will yield 20-40 µg of DNA. Other sources of DNA are accessible tissues such as placenta and buccal swabs and the quantity of DNA that can be obtained is only some hundred µg maximally. Even if the absolute mass sensitivity of the analysis is adequate, it still must be determined that the adduct can be detected in the presence of significantly greater quantities of unmodified bases. As a consequence, analytical techniques for analyzing DNA must i) be sensitive enough to detect low levels of adducts; ii) require only microgram quantities of DNA; iii) produce results that can be quantitatively related to exposure; iv) be applicable to unknown adducts that may be formed from complex mixtures, v) be able to resolve, quantify and identify adducts, vi) be inexpensive, vii) be rapid, viii) be able to analyze large number of DNA samples and ix) produce low risk to the person carrying out the procedure.

1.2.2 Common Techniques (Other Than MS)

Since the early 1960s when it was unequivocally established that DNA was a macromolecular target for chemical carcinogens, various methods have been developed for adduct analysis. These include ^{32}P-postlabeling, immunoassays, fluorescence spectroscopy and mass spectrometry. It was stated above that analytical techniques employed in DNA adduct analysis must offer powerful identification ability and excellent limits of detection (LOD). Most techniques currently favored are deficient in at least one of these requirements. As interest has grown to better understand the relationship between DNA adducts formation and cancer risk, advances have been made in both respects. Below is a comparison of different techniques applied for the analysis of DNA adducts. In choosing among the different techniques, the amount of DNA sample available and the chemical nature of the DNA adduct should be considered.

Table 1.3. Comparison of different techniques applied for the analysis of DNA adducts (Appendix, adapted and modified from De Kok et al., 2002 and Farmer et al., 2005)

Technique	Chromatography	Sensitivity (adduct/nucleotide)	Amount of DNA Needed (*ug*)	Quantification	Comments
^{32}P-Postlabeling	TLC	$1/10^{10}$	2-10	Radioactivity detection of postlabeled species. Recovery determined with standard adducted nucleotide	- Useful for screening variety of carcinogen adducts. - Applicable to analysis of complex mixtures - Danger of underestimation of adduct levels due to inefficient chromatographic recovery and phosphorylation - Interference of "indogeneous" spots
	HPLC	$1/10^{10}$	10		- Less sensitive compared to TLC (~factor 10), but a relatively large amount of DNA can be analyzed to compensate the loss of sensitivity
	CE	$1/10^{9}$	10		- Enables multiple injections of a single sample
Immunoassay	Standard ELISA	$1/10^{8}$	> 100	Calibration curve using standard adduct	- Usual high specificity for carcinogen or class of carcinogens - Interference of substances that compete with antibody recognition
	CE	$1/10^{7}$	0.5-2		- Less DNA needed but also less sensitive
Fluorescence	HPLC	$1/10^{8}$	> 100	Calibration curve using standard adduct	- Only applicable to fluorescent compounds - Information on adduct identity needed
Mass Spectrometry	GC/LC	$1/10^{9}$	> 100	Ion intensity. Calibration curve using standard adduct and stable isotope labeled internal standard	- Structural identification of adduct - Derivatization needed for GC
	CE	$1/10^{9}$	> 100		- Sample stacking used to lower detection limit
Accelerated Mass Spectrometry	HPLC	$1/10^{12}$	500	Separation of isotope ions based on mass, followed by quantification in a gas ionization detector to give an isotope ratio	- Not applicable in standard human biomonitoring studies due to use of 14C-labled substances

^{32}P-Postlabeling

The ^{32}P-postlabeling method was developed by Randerath et al. (1981) initially for the analysis of adducts of bulky aromatic compounds and for many years it was the most widely used technique in DNA adduct analysis. This method generally comprises four steps: i) Enzymatic digestion of the DNA into 3' mononucleotides. ii) Enrichment of the adduct fraction of the digest to enhance the sensitivity of the assay. This was done by solvent or solid-phase extraction (SPE), immunoaffinity chromatography, HPLC, or further enzymatic digestion. iii) Radiolabeling of the digestion products. This was done at the 5'-position of the adducts by polynucleotide kinase-mediated phosphorylation of the adduct nucleotides with [γ-^{32}P]ATP of high specific radioactivity to form [5'-^{32}P]-3'-bisphosphates. iv) Chromatographic separation of the labeled adducts by thin-layer chromatography (TLC) or HPLC, followed by detection and quantitation by measuring ^{32}P-decay.

The main advantage of ^{32}P-postlabeling is its high sensitivity as it enables the detection of 1 adduct in 10^{10} nucleotides. In addition, it requires only small quantities (2-10 μg) of DNA for the analysis, which is in contrast to LC-MS where 50-100 μg of DNA is a typical requirement. This is particularly true for the postlabeling of bulky/non-polar DNA adducts. Furthermore, coupling of the ^{32}P-postlabeling technique with HPLC separation of DNA adducts results in high specificity. ^{32}P-postlabeling is very useful for detecting DNA adducts resulting from exposure to many diverse classes of chemical carcinogens, such as PAHs, aromatic amines, HAAs, alkenylbenzene derivatives, benzene and its metabolites, styrene, mycotoxins, simple alkylating agents, unsaturated aldehydes from lipid peroxidation, pharmaceuticals, ROS, and UV radiation.

The main disadvantage of ^{32}P-postlabeling is that it does not provide structural information for identification of the DNA adduct unless standards/reference compounds are available for a chromatographic comparison. Secondly, the assay could give false-negative results due to loss of adducts or false-positive results as the result of labeling of nonnucleic acid components by polynucleotide kinase. Thirdly, the procedure can be quite labor intensive and time consuming. Lastly, it has to deal with large quantities of radioactive material requiring a dedicated radioactivity laboratory equipped with protective shielding. In quantification of a DNA adduct, the ^{32}P-postlabeling method may underestimate adduct levels due to differences in hydrolysis/incomplete DNA digestion, inefficiency of adduct labeling by polynucleotide kinase, and/or loss of adducts during enrichment and TLC stages. False-negative results occur more often with nonaromatic than aromatic adducts. False positive results also arise often from contamination of DNA with RNA which gives an overestimation of adduct levels. This is due to the fact that the standard TLC conditions are generally not adequate to resolve DNA adducts from RNA adducts.

To summarize, these are the pitfalls that ^{32}P-postlabeling faces when dealing with unknown samples in particular: i) Absolute quantitation of internal standards (IS) is difficult because conditions for digestion, enrichment (nuclease P1 reaction) and labeling (phosphorylation) cannot be optimized for each adduct. Consequently, some adducts are assayed under suboptimal conditions, especially proper standards with known adduct levels are not available for each adduct to be assayed in parallel. ii) Incomplete hydrolysis of DNA may lead to errors in quantitation, as additional spots due to the presence of adducted oligonucleotides appear in postlabeling analysis.

Immunoassays

Immunoassays have been used for some years for the analysis of DNA adducts and are becoming increasingly popular owing to their high sensitivity (can achieve 1 adduct in 10^8 nucleotides) and ease of use. Immunoassays can be direct or competitive and, since the competitive assays are more sensitive, most analyses are conducted in the competitive mode, with either monoclonal or polyclonal antibodies. The specific methods include competitive enzyme-linked immunosorbent assays (ELISA), which requires the development of antibodies to the specific adduct of interest and immune-slot blot assays that are more sensitive than ELISA. In addition, there are immunohistochemical techniques.

Using immunoassays to determine DNA adducts is based upon eliciting and characterizing polyclonal and monoclonal antisera against haptens. Different types of competitive immunoassays have been used for adduct measurement and generally quantitation is based upon competition of the antibody-antigen binding by antigen in the sample. They are most commonly performed on microtiter plates, and inhibition of binding of the specific modified DNA antiserum to a coated immunogen DNA is observed, with either biological sample DNAs or a standard curve composed of DNA modified in the same range as the biological samples. Samples are typically assayed twice on different days using triplicates for each assay. The amount of adduct in the unknown sample is obtained by comparison with the known standard. The last step is the detection of antiserum bound to a microtiter plate and can be colorimetric, fluorescent or chemiluminescent.

The main advantages of immunoassay are sensitivity, selectivity and simplicity. As a screening device, immunoassays are relatively straightforward to perform, are inexpensive, and can achieve high throughput. One person can assay about 25 samples per day and the daily reproducibility of the standard curve provides internal quality control. They also have good relative specificity because the antisera usually recognize multiple adducts from the same chemicals as well as adducts of other carcinogens in the same chemical class. Moreover, in addition to their relative simplicity, immunoassays such as ELISA are generally inexpensive.

The main disadvantages of immunoassays include: i) The requirement of relatively large amounts of DNA for maximum sensitivity. 100 µg is typical, and is not uncommon to use more than 200 µg of DNA in total if approximately 25 µg DNA are assayed per microtiter well. ii) The data obtained generally give values for relative and not absolute quantitation, unless the standard and the unknown sample are both modified by a single compound. iii) It must be possible to obtain a DNA adduct in sufficient quantities (in the range of 1%) to use as an immunogen. iv) The requirement to immunize/raise the specific antibodies from rabbits or mice (for polyclonal antibodies) and the necessity to characterize the antiserum and validate the assays. v) This approach does not work with unstable adducts. Aside from these, several polyclonal and monoclonal antibodies are available which can be used in large population studies frequently to cross react with structurally related DNA adducts formed by other carcinogens.

Fluorescence Spectroscopy

Fluorescence techniques are less often used than other methods. The sensitivity of these techniques can reach 3-10 adducts per 10^8 nucleotides. Application of fluorescence techniques, e.g., scanning synchronous fluorescence spectroscopy (SFS) takes advantage of the fluorescent properties of PAH carcinogens, carcinogen metabolites, and their adducts by simultaneously scanning excitation and emission wavelengths at a fixed wavelength difference.

The main advantage is that the assays can be performed rapidly and inexpensively, following the initial investment in the equipment. Also in some cases, the limit of detection can reach the level that is comparable to ^{32}P-postlabeling. The major limitations are: i). Prior knowledge of the chemistry of the adducts involved; ii) The requirement that the adduct has to possess a fluorophore; iii). Large quantities of DNA (100-1000 μg) are necessary to reach needed sensitivity.

In summary, all of the aforementioned analytical techniques fail to provide a definitive answer that provides both structural characterization and adequate sensitivity for quantification of biomarkers using small amounts of sample. Mass spectrometry in combination with separation methods has been viewed as a viable alternative for this purpose. In this chapter of the book, we review highlights of mass spectrometric applications to the analysis of biomarkers indicative of DNA damage. While the focus of this book's research has been on the development of HPLC-MS methodology for the analysis of DNA adducts in the form of oligonucleotides, we first present an almost historical summary of key advances in the field of DNA biomarkers in order to provide a better perspective of the problems and challenges associated with the problem. Although the significance of characterizing oligonucleotide adducts in the context of DNA damage has been known, the initial focus was on the analysis of DNA adducts as "monomeric units", i.e., nucleosides of nucleotides, in part because they presented an easier target. In the following, we review the basic strategies to DNA adduct analysis by mass spectrometry beginning with a discussion of the early approaches followed by the most recent advances in the MS analysis of oligonucleotide adducts.

1.2.3 Mass Spectrometry-Based Methods

Initial applications of mass spectrometry to the study of carcinogen related biomarkers focused on protein adducts (Bailey et al., 1987, 1990; Bryant et al., 1987, 1988; Carmella et al., 1990; Day et al., 1990; Farmer and Bailey, 1989; Farmer et al., 1986, 1988; Hecht et al., 1991; Lynch et al., 1991; Sabbioni and Neumann, 1990). There are several reasons for this. Firstly, adducted proteins are available in greater quantity (several hundred milligrams) from in vivo studies, in contrast to adducted DNA where only microgram quantities are normally available. Secondly, the earlier MS technology was based on compound separation by gas chromatography (GC) using electron impact ionization (EI) or chemical ionization (CI), which was more applicable to modified amino acids than to modified nucleotides. In addition, the generally favorable and well defined response of proteins or amino acids to MS ionization methods, especially in the positive ion detection mode, made the analysis of protein adducts simpler.

However, over the past decade, interest in the use of MS for the analysis of DNA adducts has been on the rise. DNA and DNA adducts analysis by MS has two major advantages over that of protein-adducts: 1. With its four bases, DNA is quite plain and any sequence can be analyzed with virtually the same setting of the mass spectrometer. 2. Each nucleated cell carries two copies of genomic DNA and as a consequence there is no problem with dynamic range.

Historically the role of MS in the determination of DNA adducts had been limited to providing information for the identification of new DNA adducts, or for the structural characterization of DNA adduct standards that have been utilized to determine adduct levels by other detection methods, such as ^{32}P-postlabeling. However, MS has gained quite a momentum in recent years largely owing to the development of improved chromatographic interfaces and on-going technological advances in ionization methods, in particular ESI as well as ion transmission and detection. These improvements have led to significant improvements in sensitivity and efficiency compare to a decade ago. In contrast, the sensitivity of ^{32}P-postlabeling has not improved any further while, at least for targeted analysis, the limit of detection of MS can reach 1 adduct in 10^9 nucleotides and the DNA quantity used for such adduct analysis can be as low as ca. 10 μg. Therefore the power of MS to achieve qualitative and quantitative analyses of human DNA adducts has increased greatly and it is likely to become the method of choice for DNA adduct analysis in the not too distant future.

In fact, the growing use of MS for the analysis of DNA adducts can be gauged by the increasing number of publications between year 1996 and 2006 (SciFinder). Traditional ionization methods such as EI, CI found limited use but the development of fast atom bombardment (FAB) offered new opportunities for direct analysis of DNA adducts albeit with somewhat disappointing sensitivity. MALDI accounted for a small portion of the DNA adduct analysis as it is an off-line technique. Gas chromatography in combination with negative ion chemical ionization (GC-NICI-MS) has provided significant improvement in sensitivity but its usage has gradually diminished with the steady growth of LC-MS following the development of electrospray ionization techniques. These trends in MS applications to DNA adducts analysis are illustrated in **Figure 1.2** below. In this regard, it is also important to identify the leading journals in which these publications have been appearing and this is shown in **Figure 1.3**.

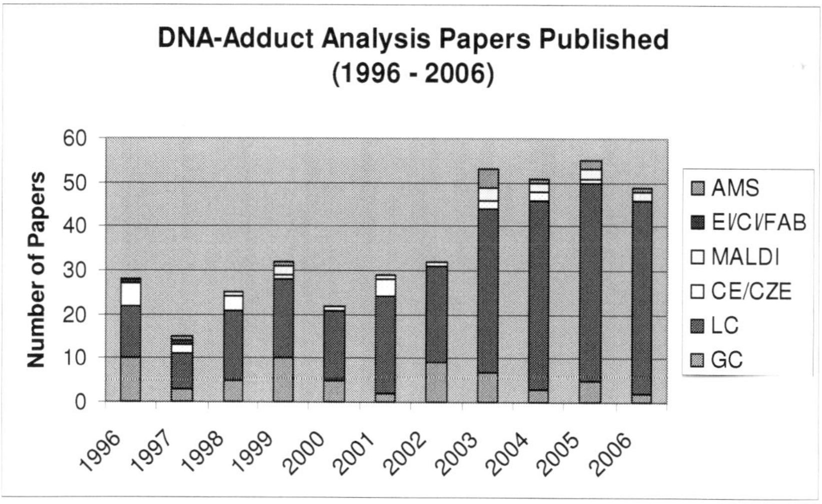

Figure 1.2. DNA adducts analysis papers published between year 1996 and 2006, searched through SciFinder, categorized by the techniques used.

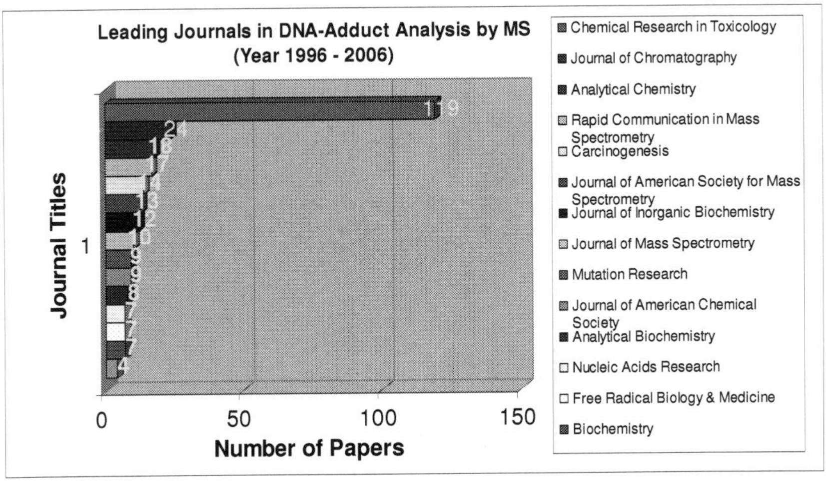

Figure 1.3. Leading journals with most publications in DNA adducts analysis between year 1996 and 2006, searched through SciFinder.

The fundamental approach to the analysis of DNA adducts by MS methods is to some extent analogous to that used in ^{32}P-postlabeling. Specifically, using as an illustration **Figure 1.4** which shows a short segment of DNA with a PAH carcinogen attached to a nucleobase, the first step in the analysis involves enzymatic hydrolysis of the DNA polymer (or oligomer) into monomeric units, either nucleosides or nucleotides. For purposes of the mass spectrometric analysis, these monomeric units can be viewed as consisting of effectively three "blocks": the base (typically **G** and occasionally **A**), the deoxyribose (or deoxyribose monophosphate) and the carcinogen (shown in this figure in red). Alternatively, if interested in identifying the sequence context of adduct attachment (which is the main objective of this review), enzymes capable of hydrolyzing the DNA into oligonucleotide adducts may be utilized. In either case, extensive sample pretreatment needs to be conducted in order to remove the bulk of the unmodified DNA and reconstitute the adducts into a final volume enriched in the analytes of interest. The specific sample cleanup procedures may differ depending on the type of adducts involved and will not be discussed any further in this review.

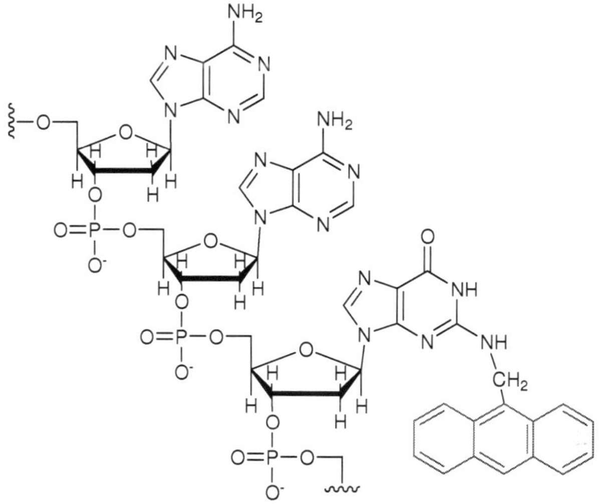

Figure 1.4. A segment of DNA showing formation of a PAH adduct on the exocyclic N^2-position of deoxyguanosine.

In line with the above considerations, the approaches taken for the analysis of DNA adducts, whether monomers or oligomers, by mass spectrometry have varied over the years and, as might be expected, they followed the technological developments associated with MS ionization and ion separation methods. While this review is intended to focus primarily on developments related to the analysis of oligonucleotide adducts, highlights of MS applications for the analysis of monomeric adducts have been included in order to provide the broader perspective of the field.

1.2.3.1 Gas Chromatography – Mass Spectrometry (GC-MS)

For many years and before the advent of LC-MS, GC-MS was the main MS method available for the analysis of DNA adducts. However, since analyte volatility is an essential requirement for GC-MS, this precluded the analysis of intact DNA adducts regardless of whether they were in the nucleoside or nucleotide form. For small adducts, e.g., oxidation or etheno adducts, enzymatically generated adducts would be deglycosylated and the released modified nucleobase analyzed by converting to a suitable volatile derivative.

Major improvements in sensitivity have been realized via the use of electrophoric derivatives in conjunction with negative ion chemical ionization (NICI), originally introduced by Hunt and Crow (1978). A commonly used derivatization method developed by Abdel-Baky and Giese (1991) and Giese et al. (1996) involves replacement of active hydrogens by pentafluorobenzylation as shown in **Figure 1.5**. This leads to products that usually undergo facile, dissociative electron capture to form an analyte-characteristic anion. A detailed review of the use of electrophoric derivatives in the analysis of DNA adducts by GC-MS has been published by Giese (2000).

Some specific examples of the application of GC-MS include the work of Chaudary et al. (1994), who detected endogenous adducts in human liver by GC-EC-NICI-MS. Moller and

Hofer (1997) used GC-MS, HPLC-EC, and ^{32}P-postlabeling to study the correlation of increased level of 8-hydroxy-2'-deoxyguanosine (8-OH-dG) and diseases like diabetes, infections and cystic fibrosis. Also, Topinka et al. (1998) revealed the presence of PAH and nitro-PAH derived DNA adducts in mammalian cells in culture by GC-MS. GC-MS was also used to quantify M1G and their equivalent DNA adducts derived from the reaction of malondialdehyde (MDA) and DNA (Otteneder et al., 2002) while more recently Turesky et al. (2003) used GC-NICI-MS to identify elevated ABP DNA adducts caused by hair dye usage. A good example of such applications is that of the Giese group (Wu et al., 1999) on the analysis of endogenous and exogenous N7-HEG dG adducts by Gas Chromatography Electron Capture Negative Ion Chemical Ionization High Resolution Mass Spectrometry (GC-EC-NICI-HRMS) and comparison of their results with those obtained by HPLC-MS.

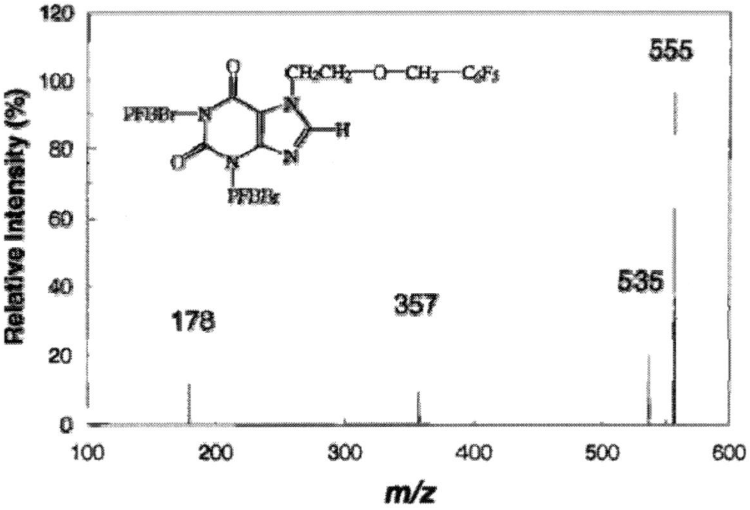

Figure 1.5. Full-scan spectrum of the derivatized *N*7-hydroxyethyl guanine (N7-HEG. The spectrum shows no molecular ion peak, but the most abundant fragment is *m/z* 555 corresponding to [M – 181]$^-$ (Wu et al., 1999).

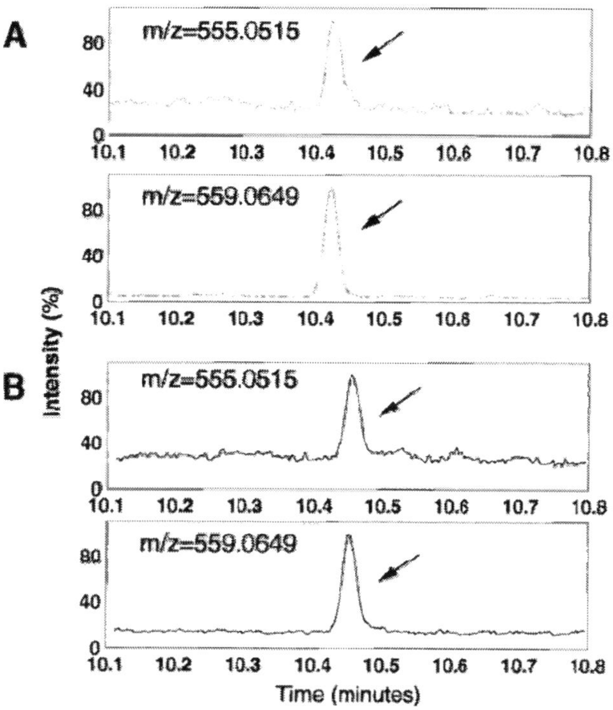

Figure 1.6. Representative chromatograms of (A) spleen DNA of a control rat and (B) brain DNA of a rat exposed to 3000 ppm ethylene for 6 h/day for 5 days (Wu et al., 1999).

As noted above, since it is generally impossible to analyze an intact adduct by GC-MS, in most cases a typical strategy involves removal of the sugar in order to work with the adducted free nucleobase. However, working with the free base may introduce potential RNA interferences. Alternatively, especially for bulky PAH adducts, it is possible to combine the two steps of chemical transformation and electrophoric derivatization as illustrated for the analysis of the deoxyguanosine adduct of BaP. In this indirect approach, the analyte was transformed by mild acid hydrolysis and oxidation to the pyrene-1,2-dicarboxylic acid which was then pentafluorobenzylated for detection by GC-EC-NICI-MS (**Figure 1.7**) (Allam, et al., 1993).

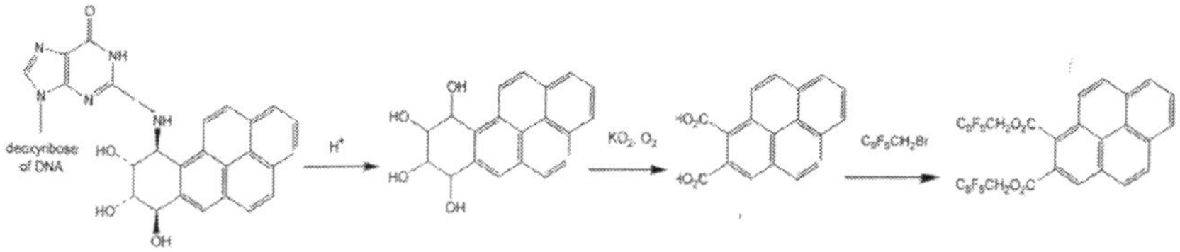

Figure 1.7. Chemical transformation and electrophoric derivatization of a benzo[*a*]pyrene diolepoxide (BPDE) DNA adduct (Allam, et al., 1993).

In addition to limitations imposed by the poor volatility of DNA adducts, it should be realized that chemical derivatization methods such as silylation and electrophore labeling can bring their own sets of problems, including the production of artifacts, interferences and sample to sample variability in derivatization (Kresbach et al., 1989). A classical case is that of adducts formed via oxidation processes. Specifically, special precautions have to be taken for the analysis of 8-oxo-deoxyguanosine (8-oxo-dG) because the multiple steps associated with sample preparation for GC-MS analysis end up producing oxidation products of dG which results in overestimation of the 8-oxo-dG content. As a result, the use of GC-MS for the analysis of carcinogen-DNA adducts in biological matrices is gradually being replaced by HPLC-MS as the latter has now reached a level of maturity.

1.2.3.2 Liquid Chromatography – Mass Spectrometry (HPLC-MS)

The clear advantage of HPLC over GC is the ability to work with thermally labile compounds and, when viewed in the context of DNA adducts, this translates into the ability to handle intact DNA adducts either in their nucleoside, nucleotide as well as oligonucleotide forms. We present here highlights of HPLC-MS applications to the analysis of DNA adducts in that sequence. Straub and Burlingame as early as 1981 wrote an extensive review on the use of mass spectrometry for the analysis of xenobiotic-modified nucleic acids (Straub and Burlingame, 1981). At that time DNA adducts were often first separated by LC or CE and analyzed by off-line mass spectrometric methods. Subsequent reviews have focused on the analysis of DNA adducts by capillary methods, both CE and LC, coupled to mass spectrometry (Andrews et al., 1999; Apruzzese and Vouros, 1998). Turesky and Vouros (2004) reviewed the principal methods for synthesis and characterization of DNA adducts and summarized the detection of the heterocyclic aromatic amine adducts in experimental animals and humans. The features of different analytical methods including LC-MS, AMS (Accelerated Mass Spectromerty) and ^{32}P-postlabeling as applied to the measurement of heterocyclic aromatic amine (HAA-DNA) adducts in vitro and in vivo were discussed. Other important reviews include those of Bohr (1991), Esaka et al. (2003), Esmans et al. (1998), Garner (1998), Phillips et al. (2000), De Kok (2002), Doerge et al. (2002), Stiborova et al. (2004), and Farmer et al. (2005). Perhaps the most detailed and comprehensive coverage of the most recent literature can be found in a review by Singh et al. (2006).

Analysis of DNA Adducts in the Mononucleoside Form

The introduction of electrospray ionization in combination with tandem mass spectrometry opened numerous new opportunities for the analysis of DNA adducts. In 1995 Chaudhary et al. published the first paper using LC-ESI-MS methods for nucleoside adducts analysis (Chaudhary et al., 1995). Referring back to the fundamental structural features of a DNA adduct (**Figure 1.4**), it is almost fortuitous that the MS/MS behavior makes them ideally suited for low level detection and quantification. Notably, a favorable collision induced dissociation (CID) of the $[M+H]^+$ ions of adducts in positive ESI ionization involves the loss of the deoxyribose block (116 amu). A low intensity fragment at m/z 117 amu is also observed occasionally and corresponds to the protonated 2'-deoxyribose "block". The facile cleavage of the glycosidic bond lends itself for trace level detection and quantification with selective reaction monitoring (SRM) using the transition $[M+H]^+ \rightarrow [M+H-116]^+$ in triple quadrupole mass spectrometers. In addition, since the fragmentation appears to be common to almost all examples of nucleoside adducts reported in the literature, operation of the tandem MS in the constant neutral loss (CNL) mode allows selective screening of complex chromatographic mixtures for the presence of adducts. Although the sensitivity of CNL is much lower than SRM, its selectivity does yield information for the determination of different DNA adducts in mixtures.

Numerous applications of LC-MS/MS are based on this characteristic behavior of nucleoside adducts. In what was perhaps one of the earliest applications of capillary LC-MS/MS to the analysis of DNA adducts, Wolf and Vouros (1994) capitalized on this process to confirm the presence of deoxyguanosine adducts of N-acetylaminofluorene (AAF) in calf thymus DNA (ctDNA) reacted with N-acetoxy-2-acetylaminofluorene (AAAF). Moreover, a CNL screen of the same mixture allowed the identification of deoxyadenosine adducts whose occurrence had not been previously verified.

The commonality of this fragmentation is illustrated in the example of **Figure 1.8** which shows the CID spectra of two isomeric adducts, dG-C8-IQ and dG-N^2-IQ, produced from the heterocyclic aromatic amine 2-amino-3-methylimidazo[4,5-*f*]quinoline (IQ), a carcinogen generated during the cooking of beef or fish. Both spectra exhibit the characteristic loss of 116 amu while an additional major peak at m/z 331 (loss of NH_3 from m/z 348) is distinctly more intense in the spectrum of the dG-N^2-isomer (Soglia, 2001). These general trends in the fragmentation patterns of DNA adducts are further illustrated in **Figure 1.9**, which shows the MS/MS spectra of a series of nucleoside, nucleotide and depurinated adducts (Singh et al., 2003).

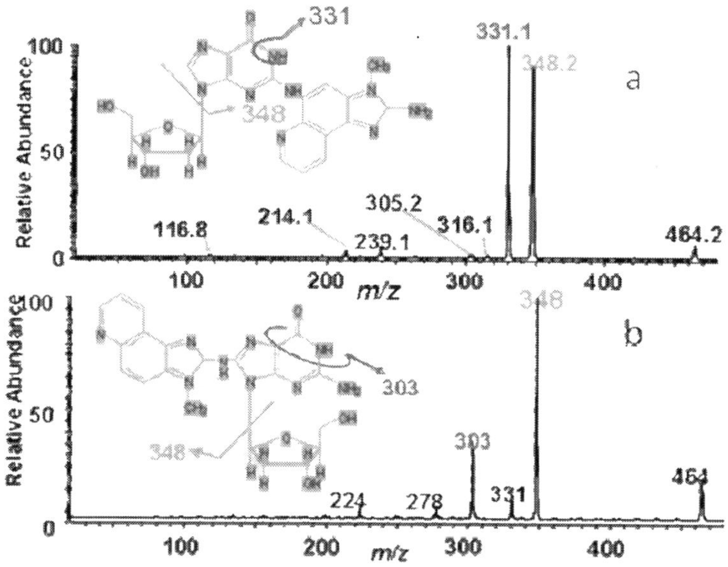

Figure 1.8. Product ion spectra of (a) dG-C8-IQ and (b) dG-N^2-IQ (Soglia et al., 2001).

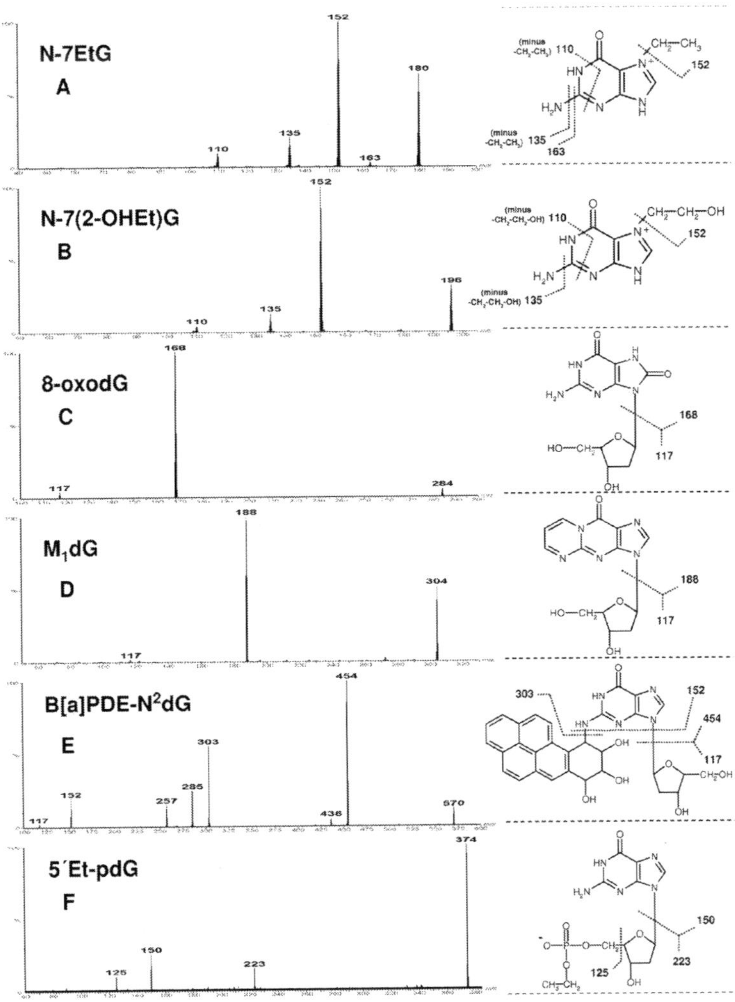

Figure 1.9. Typical ESI-MS/MS CID product ion spectra obtained using a Micromass Quattro Ultima Pt. tandem quadrupole mass spectrometer (Micromass, Waters Ltd., Manchester, UK) for different guanine $[M+H]^+$ precursor ions of nucleic acid base, 2'-deoxynucleoside and $[M–H]^-$ precursor ion of 2'-deoxynucleotide DNA adducts (adapted from Singh et al., 2003).

Capillary LC and Nano LC-ESI-MS

The introduction of capillary LC (CapLC) columns has opened new opportunities for the analysis of DNA adducts at the trace level and the increased sensitivity brought LC-MS technology within the range of human applications. The improved mass sensitivity was significant when compared to normal bore LC-ESI-MS due to improved ionization efficiency at the lower flow-rates. Abian et al. (1999) reviewed the practical considerations and technical aspects that need to be taken into account for the implementation of capLC with ESI-MS. The benefits derived from the use of capillary chromatography and nano (or micro) ESI systems, specifically as they pertain to DNA adducts have been discussed by the Esmans group. Applications to the analysis of a wide variety of adducts have been reported by this group including those of estrogen metabolites and other mustards (Embrechts et al., 2001, 2003; Hoes et al., 2000; Van den Driessche et al., 2004a, 2004b; Vanhoutte et al., 1997;).

The special features offered by capillary LC have been extensively exploited by the Vouros group and considerable improvements have been achieved over the past several years by exploring the use of progressively smaller diameter columns, with a 75 μm i.d. column providing a reasonable compromise in terms of sample capacity (loadability), and sensitivity. Some representative results achieved for the analysis of nucleoside adducts using capillary columns are discussed next.

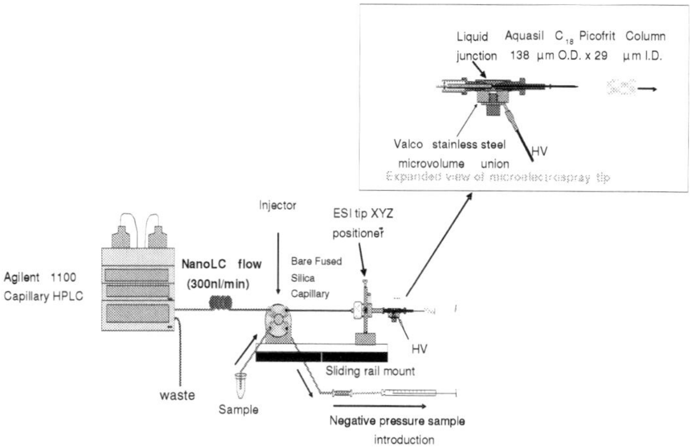

Figure 1.10. Schematic of capillary liquid chromatography system in line with Finnigan TSQ 700 mass spectrometer (adapted from Riciki et al., 2006).

A capillary LC-MS interface designed for optimal sensitivity is shown in **Figure 1.10** in which the electrospray tip itself has been modified to also serve as the column. Samples are deposited directly onto the top of the column for preconcentration and eluted with a stronger solvent. The absence of extra column effects results in better resolution and sharper chromatographic peaks.

The configuration shown in **Figure 1.10** was used in a dose response study in which rats were administered the carcinogen IQ and the levels of dG-C8-IQ in liver were determined by LC-MS/MS (Soglia et al., 2001). Chromatograms representing the detection of the adduct at each level of administration are shown in **Figure 1.11**. The profiles of the isotopically labeled internal standard are included in each panel. Adduct levels ranged from as little as 3.5 modifications per 10^8 normal bases at the lowest dose (0.05 mg/kg) to as high as 38 per 10^8 at the highest dose (10 mg/kg) with RSD (relative standard deviation) of less than 20%.

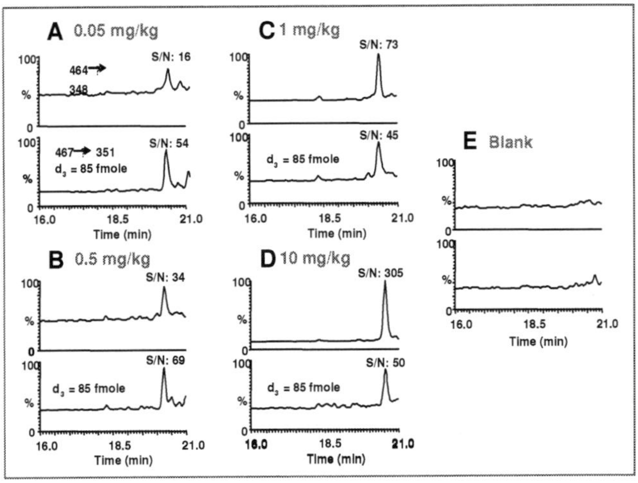

Figure 1.11. Selected chromatograms of liver DNA from rats dosed with IQ at different levels (adapted from Soglia et al., 2001).

Impressive results have been obtained for deoxyguanosine adducts of 4-ABP using the improved interface of **Figure 1.10**. **Figure 1.12** shows a chromatogram illustrating detection of 3 fg and 7.7 fg injected on-column for dG-C8-ABP and dG-C8-ABP-D_9 the unlabeled and labeled analogs of the adduct. The standard curve for this analyte generated from 300µg of DNA shows detection limits of 0.55 fmol as illustrated in **Figure 1.13**.

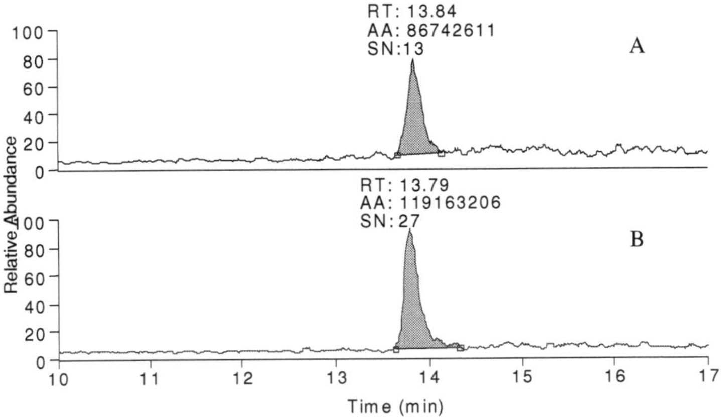

Figure 1.12. (A) Extracted ion chromatogram (435 → 319) of undigested 6 fg/L (3 fg on-column) solution of dG-C8-ABP; (B) Extracted ion chromatogram (444 → 328) of undigested 15 fg/µL (7.5 fg on-column) solution of dG-C8-ABP-D_9 (adapted from Riciki et al., 2006).

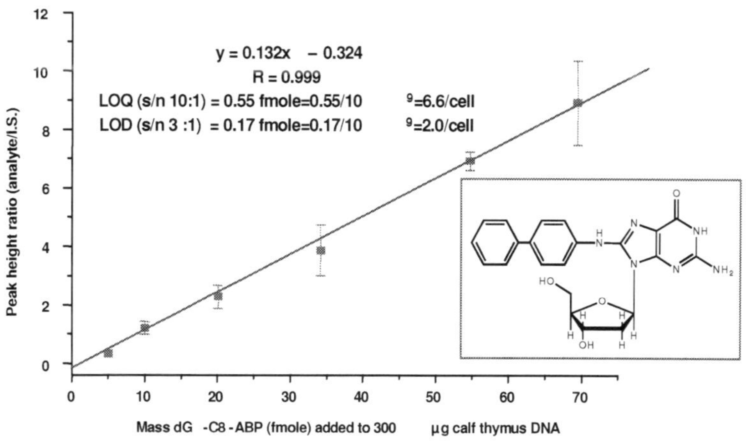

Note: 20 fmole adducts in 300 µg DNA correspond to 2/10 8nt and 240/cell.

Figure 1.13. dG-C8-ABP Standard Curve (adapted from Riciki et al., 2006).

The method was then used to measure the steady state levels of the same adduct in the human TK6 lymphoblastoid cell line as a function of dose and time after exposure to the metabolite of 4-ABP (Ricicki et al., 2006). The method provided a novel, validated method for the analysis of dG-C8-ABP DNA adducts by nanoLC-μESI-MS and was utilized for the quantification of the adducts present in human TK6 cells from both a dose and time response study. **Figure 1.14** shows representative chromatograms based on variation of dose levels. Panel 1 shows extracted ion chromatograms (EIC) (*m/z* 435 → 319, dG-C8-ABP) from the analyses of cells harvested at 6 hours, while the chromatograms in Panel 2 resulted from the analyses of cells harvested at 27 hours. As labeled in **Figure 1.14**, chromatogram **a** represents the analysis of the blank human TK6 cells used as the control. There was a small peak in the blank that was consistently observed throughout all the analyses of human cells. The reproducible retention time of 15.29 minutes demonstrated that this impurity did not interfere with and was not related to the analyte of interest, dG-C8-ABP, which eluted at approximately 14.0 minutes. The quantification results were correlated to cell toxicity, induced mutation at the TK and HPRT loci, and gene expression profiling through microarray analysis.

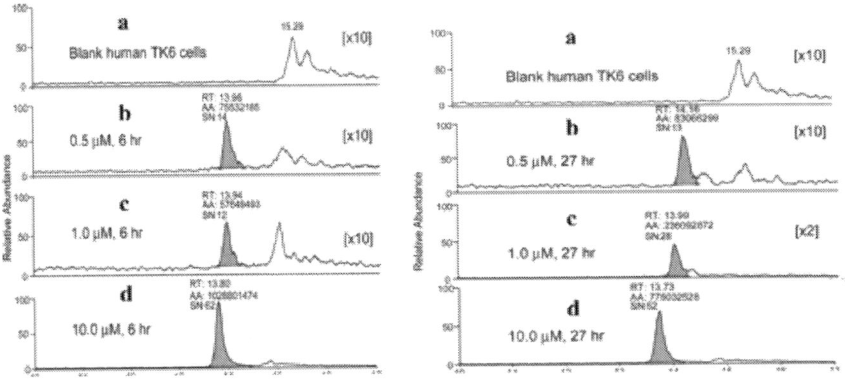

Figure 1.14. Extracted ion chromatograms of dG-C8-ABP (435→319) from the dose response study. (a) in 1 and 2 is the chromatogram from the analysis of blank human TK6 Cells. Panel (1) chromatograms from harvesting 6 hours after dosing; Panel (2) 27 hours after dosing. (b) 0.5 M dosing (c) 1.0 M dosing (d) 10.0 M dosing (adapted from Ricicki et al., 2006).

Analysis of DNA Adducts in the Mononucleotide Form

As noted above, DNA adducts are analyzed by LC-ESI/MS primarily in the form of nucleosides and in the positive ion mode. Detection of nucleoside adducts in positive ESI is typically at least an order of magnitude more sensitive than detection of the corresponding nucleotide adduct which has to be analyzed in the negative ion mode due to the presence of the phosphate group. Moreover, nucleotide adducts are often too hydrophilic to be separated by the most popular reversed phase HPLC (RP-HPLC) methods and, as a result, best separations are achieved using ion pairing methods. The high electrolyte content of the mobile phase, however, introduces ion suppression effects in the electrospray ionization which impacts negatively on the sensitivity of the assays. There are exceptions such as adducts of thymidine glycol and 5-hydroxy-2'-deoxyuridine which are actually detected with greater sensitivity in negative ESI compared with positive ESI and undergo cleavage of the glycosidic bond following collision induced dissociation (CID). Nucleotides may also be first separated by anion-exchange SPE followed by enzymatic dephosphorylation and LC-MS analysis of the resultant nucleosides. The addition of extra cleanup steps, however, is not encouraged as it may lead to sample losses particularly when dealing with trace levels of adducts. In order to overcome these problems, attention has been turned to the use of capillary electrophoresis (CE-MS), a brief discussion of which follows later.

Despite the above limitations, analysis of DNA adducts, whether monomeric or oligomeric, in the nucleotide form offers the additional advantage of being able to detect alkylation to the phosphate group, i.e., damage to the DNA backbone, which would be missed in the analysis of the dephosphorylated analog. An excellent example of particular current interest is the identification of phosphate adducts of phenylglycidyl ether in studies conducted by the Esmans group in the 1990's on bisphenol A (BPA) and related analogs. BPA was first synthesized in 1891 and BPA products have been on the market for over 50 years as more than 2 billion lb are used annually in the US (Erickson, 2008). BPA is a principal reactant in the preparation of polycarbonate (PC) plastics and is also widely used to make epoxy-based polymer resins or to strengthen plastics as plastic additives, lacquers, or surfactants. BPA and analogs show up in canned food, beverages and utensils (Kuo and Ding, 2004; Maragou et al. 2006; Shao et al., 2005; Uematsu et al., 2001). Reusable baby bottles and other housewares are often made of PC. In consumer products, BPA analogs can also be found in compact discs, eye glasses, bicycle helmets and automotive parts. Their presence is found in environmental samples as well, namely wastewaters and sewage (Markham et al., 1998). BPA upon dental treatment could be released into saliva from denture which contains resins (Manabe et al., 2000).

BPA has been studied for 40 years and there have always been controversial reports regarding its benefits and risks. In recent years much concern has been raised regarding the health risks of low-level exposure to BPA despite two opposing views expressed by its advocates and opponents (Hillem, 2007b; Schultz and Moore, 2008). Environmental exposure to these phenolic compounds which can act like the hormone estrogen has been associated with adverse reproductive and developmental effects in wildlife and humans which potentially leads to breast cancer and other hormonally-mediated health outcomes (Tominaga et al., 2006). Numerous studies published in the past decades have linked BPA exposure to increased rates of prostate and breast cancer, reproductive abnormalities, decreased sperm count, accelerated puberty in females, neurological effects similar to attention deficit hyperactivity disorder (ADHD), diabetes, and obesity in animal studies (Hillem, 2007a). It was also linked to heart disease. 92% of Americans age 6 or older test positive for BPA according to the data from the Center for Disease Control and Prevention (CDC). Bisphenol A can be present in urine, serum and plasma samples (Kuklenyik et al., 2003; Sambe et al., 2005; Tsukioka et al., 2003).

A broad range of DNA adducts of bisphenol A diglicydyl ether (BPADGE) have been identified by LC-MS/MS in in vitro and in vivo studies conducted by the Esmans group (Vanhoutte et al., 1997). The presence of the prominent ion at m/z 195 (the deoxyribose phosphate ion) in the MS/MS spectra indicates adduct formation through base modification and the transition $[M-H]^-$ → m/z 195 can be used to recognize those adducts formed by alkylation on the nucleobase [Panels (d)-(f) in **Figure 1.15**]. On the other hand, phosphate alkylation results in alternative fragmentation pathways and spectra void of m/z 195 fragments [Panels (a)-(c) in **Figure 1.15**]. The structures of two isomeric, phosphate and base alkylated adducts identified in these studies are shown in **Figure 1.16**.

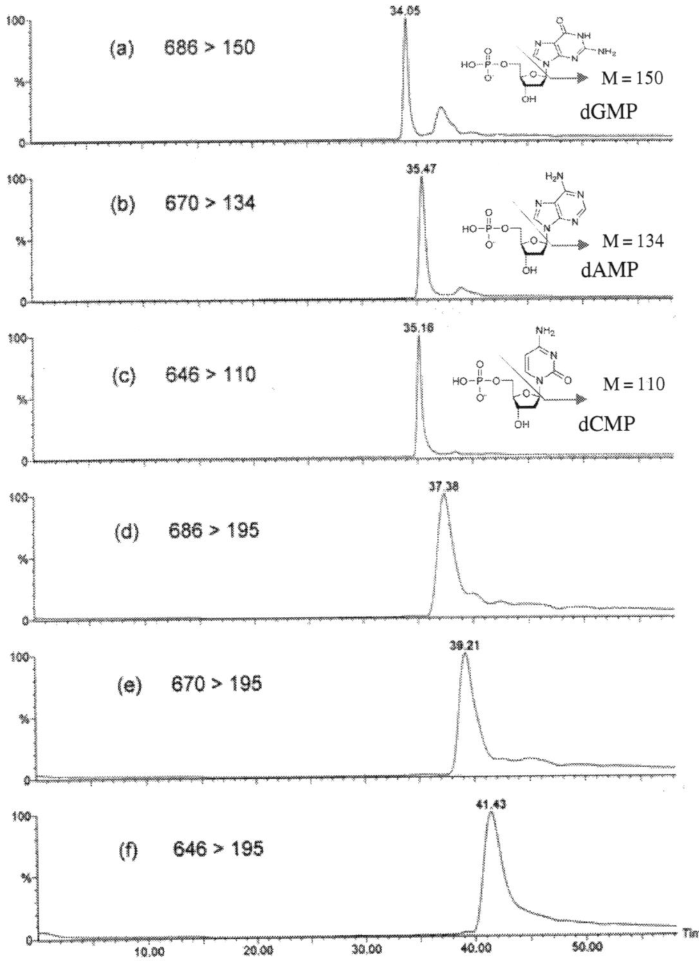

Figure 1.15. Injection of 1 μL of the dAMP/dCMP/dGMP/BPADGE in vitro reaction mixture onto the NanoFlow ES LC/MS column-switching system under MRM conditions: (a-c) phosphate-alkylated adducts (dGMP/BPADGE, dAMP/BPADGE and dCMP/BPADGE, respectively); (d-f) base-alkylated adducts (dGMP/BPADGE, dAMP/BPADGE and dCMP/BPADGE, respectively (Vanhoutte et al., 1997).

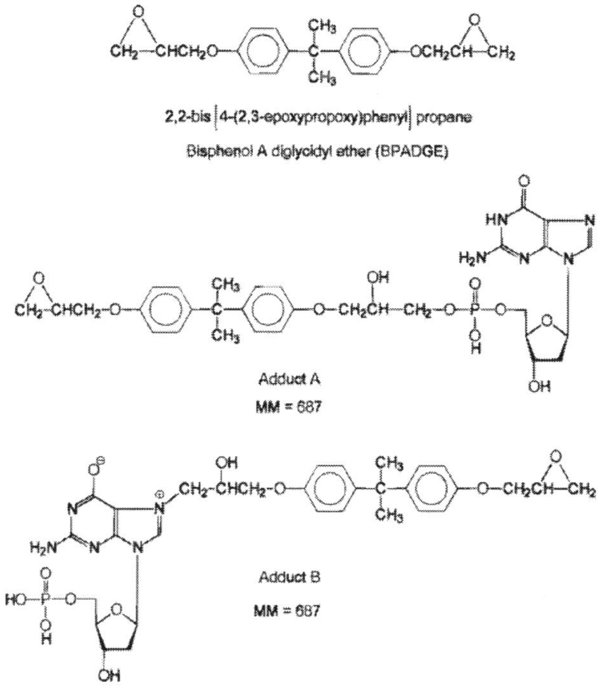

Figure 1.16. Structures of BPADGE, phosphate-alkylated dGMP/BPADGE (adduct **A**) and base-alkylated dGMP/BPADGE (adduct **B**) (Vanhoutte et al., 1997).

Analysis of DNA Adducts in the Oligonucleotide Form

Significance

Analysis of DNA adducts as monomeric species such as nucleosides or nucleotides fails to provide information about the adduct(s) location within the DNA sequence. However, it has been observed that reaction of carcinogens with DNA often exhibits sequence selectivity, the recognition of which may be important in order to understand their mutagenic activities (Belguise-Valladier and Fuchs, 1991; Hartley et al., 1992; Koehl et al., 1989b; Kohn et al., 1987; Mattes et al., 1986). Following exposure to these genotoxic carcinogens, characteristic mutation spectra then result which show a degree of site-selectivity in production of mutations. Sequence selectivity of adduct formation is manifested in a number of oncogenes. For example, mutational hotspots in the *p53* gene are particularly frequent in domains that include CpG dinucleotides (Hemminki et al., 2000). This gene has been under extensive scrutiny and shows strong evidence linking adducts, mutation and cancer. The specific mutations appear to be present in liver and lung cancers in particular. There are at least three separate databases on *p53* mutations, one in IARC with over 10,000 mutations. The predominant hotspots often correlate with major DNA adducts formation (Hartley et al., 1992; Kohn et al., 1987; Mattes et al., 1986). The mutation efficiencies of different adducts are different, ranging from barely mutagenic to those that frequently cause mutations. It is also claimed that the potency of an adduct is related to its base sequence context and the latter also influences both the types and the frequencies of ensuing mutation. It has been determined that several factors affect the ability of an adduct to induce mutation. These include orientation of the modifying group especially with respect to the base-pairing face, size and conformation of the group, nature of polymerase, sequence context and rearrangement of the adducted base due to tautomerization, rotation and wobbling. Many DNA repair deficiency syndromes involve vast increases in risk of cancer relevant to DNA adducts. The sequence selectivity or in other words, the site specificity found in a few number of mutants is striking yet unexplained (Belguise-Valladier and Fuchs, 1991; Fuchs, 1984; Hartley et al., 1992; Koehl et al., 1989b; Kohn et al., 1987; Mattes et al., 1986; Rill and Marsch, 1990; Tebbs and Romano, 1994).

In view of these considerations, the assignment of the sites of modification by different carcinogens within a target sequence is a very important topic in cancer etiology since isomeric adducts may possess different mutagenic activities due to different biological responses to the presence of the adducts. Therefore a critical part in the analysis of DNA adducts is to develop a method to determine the site and frequency of modification on the DNA strands in order to better understand the relationship between the chemical behavior of different carcinogens and mutation hot spots. In order to accomplish this goal, it is necessary to have a fast and accurate methodology for structure characterization of oligonucleotide-carcinogen adducts from *in vitro* and *in vivo* sources using on-line separation coupled with tandem MS. Although HPLC is presently the preferred method for oligonucleotide analysis, other separation methods, most notably capillary zone electrophoresis (CZE) and capillary electrochromatography (CEC) have also been utilized.

After a brief summary of the principles associated with the collision induced dissociation (CID) mass spectra of oligonucleotides, the present status and key developments in the characterization of oligonucleotide adducts by electrophoretic and chromatographic separation methods coupled to MS/MS are summarized. Examples of the most pertinent applications of these techniques are presented in order to identify the analytical challenges involved and bring into perspective the goals of the research undertaken in this work.

Oligonucleotide Sequencing by MS/MS

The ability of ESI-MS to form multiply charged ions in combination with improved mass range in modern MS instruments has made possible the detection of increasingly larger oligonucleotide fragments. In practice, because oligonucleotides have an acidic phosphodiester backbone, oligonucleotide adducts are usually analyzed in ESI negative ion mode more than in the positive ion mode. Negative ESI requires careful adjustment of sample conditions and instrument tuning to obtain sensitivity comparable to positive ESI. MS analysis can provide a basis for detecting both length and sequence variations of oligonucleotides, yet this is complicated by multiply charged ions and metal adduct ions. In ESI a mass spectrum may contain an envelope of peaks which corresponds to ions with various charge states with a Gaussian like distribution. To further complicate the picture, metal adduct ions are formed due to the affinity of the polyanionic backbone for ubiquitous cations, such as sodium and potassium. These multiply charged ions and the Na+ and K+ adduct ions (often referred to as quasi-molecular ions) contribute their own signals adding a further degree of complexity to the spectral pattern. As a result, the ability to characterize mixtures of oligonucleotides may be compromised. To overcome this problem, many different methods are used for sample desalting and purification, so as to exchange the nonvolatile cations (Na+, K+, etc.) with protons (H+) or ammonium ions (NH_4+).

Sequence determination of DNA was made possible by MS/MS based on the gas-phase collision induced dissociation (CID) of multiply charged oligonucleotide ions. McLuckey , Habibi-Goudarzi et al. (1992, 1993, 1994) developed a system for sequence determination of oligonucleotides based on well defined fragmentation pathways in their product-ion spectra. Starting from the 5'-terminus (3'-terminus), a-, b-, c-, and d- fragment ions (z-, y-, x-, and w- fragment ions) are generated upon fragmentation at the different positions of the phosphodiester group as illustrated in **Figure 1.17**.

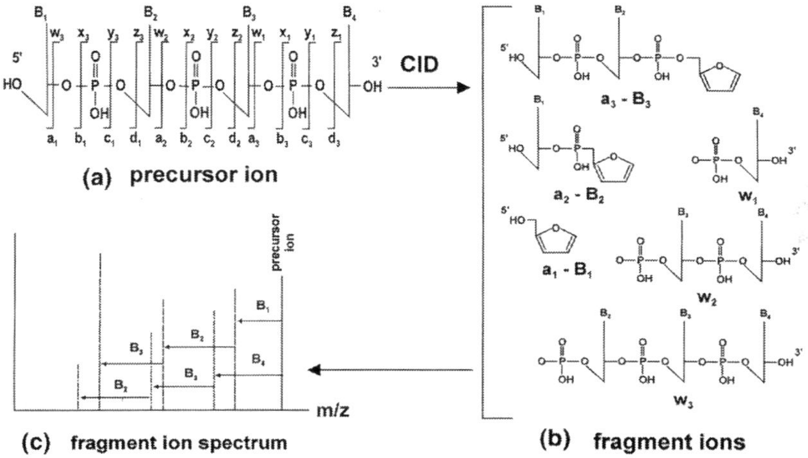

Figure 1.17. Scheme for the sequencing of DNA by MS/MS upon CID of a multiply charged precursor ion (charges are omitted in the structures.) (adapted from McLuckey et al., 1992).

The first fragmentation step is usually the elimination of a nucleobase, followed by cleavage of the 3'-phosphodiester bond of the nucleotide that suffered the base loss. Under low-energy CID conditions, fragments of the a- and w-B ions are the predominant types of species, where -B indicates the loss of the nucleobase. The resulting series of 5'- and 3'- ions (**Figure 1.17b**) have characteristic mass differences, from which the sequence can be deduced (**Figure 1.17c**).

Because of the large number of fragment ions present in the MS/MS spectra of oligonucleotides, manual spectral interpretation is difficult, time-consuming, and prone to error. As a result, efforts have been made to simplify the process of relating MS fragmentation pattern with structure. A computer-based algorithm for the sequencing of oligonucleotides of completely unknown sequence was developed by Ni et al. (1996) to automatically derive sequence information from the MS/MS spectra. The algorithm works by extending from the 5' (a-B ions) and 3' (w- ions) ends ion series that encode the complete DNA sequence. Mass ladders are identified by sequentially adding each of the four possible nucleotide masses, and by searching the spectrum for the best match of the expected ions. Rozenski and McCloskey (1999) also developed a novel approach for nearest-neighbor determination based on the analysis of fragment ions of the nucleic acid formed in the ionization region of the mass spectrometer along with the fragment ions from MS/MS. They later demonstrated the ab initio determination of unknown oligonucleotide sequences at approximately the 12-mer level and below and the approach was termed simple oligonucleotide sequencer (SOS) (Rozenski and McCloskey, 2002). However, this method cannot be extended to longer oligonucleotide sequences because usually several missing fragments in the series of a-B or w ions prevent the successful passage through the whole sequence.

Capillary Electrophoresis MS (CE-MS)

A comprehensive review on CE-MS interfacing methods has been written by Ding and Vouros (1999) while the use of CZE-MS for the analysis of PAH-DNA adducts was reviewed by Marzilli et al. (2001). Willems et al. (2005) have addressed key aspects of the analysis of nucleic acid constituents by on-line CE-MS and devoted a large section to field-oriented applications.

Because of their negative charge, nucleotide adducts are amenable to analysis by electrophoretic methods. Since the phosphate group is negatively charged over a wide pH range, electroosmotic flow conditions in traditional capillary zone electrophoresis (CZE) provide an ideal mode for their separation. Compared to HPLC, another advantage of CZE is the ability to work with small sample quantities and in cases where one is sample limited, it is possible to perform multiple measurements from a single sample. In addition, sample throughput may be considerably higher because there is no dead time for column re-equilibration. Despite these inherent advantages, a clear limitation of CE/CZE is the relatively small size of sampling volume (1-10 nL) that can be introduced into the CE-MS system. Thus, although the mass detection limit in CE can be very low, the concentration detection limit may be as high as 10^{-6} M or greater depending on the detection method.

The concentration detection limits in CZE can be improved by on-line enrichment using sample stacking techniques (Chien and Burgi, 1992). As shown by Wolf and Vouros (1995), the method resulted in 1000-fold improvement in the CZE-MS analysis of nucleotide adducts of AAF and injection volumes as high as 2-3 μL was possible. Deforce and Van den Eeckhout (2001) described the methodology developed in the analysis of DNA damage by CZE-ESI-MS and a sample stacking procedure that was able to improve the detection. Willems et al. (2002) analyzed BPDE DNA adducts by CE-MS in conjunction with sample stacking. Gennaro et al. (2004) developed a sample preparation approach which involved a variation on the digestion procedure, in combination with the use of metal affinity ZipTips to achieve more efficient cleanup of the BPDE-DNA adducts formed in vitro and subjected the bulky hydrophobic adducts to CE/MS analysis.

An interesting example (Barry et al., 1996b) of the value of sample stacking is illustrated in **Figures 1.19**, **1.20** and **1.21**, which show the results from the CZE-MS analysis of a calf thymus DNA sample exposed to BPDE, a major metabolite of BaP which is the prototypic carcinogen in the class of PAHs. The CZE-MS electropherogram of the solution obtained by hydrodynamic injection of approximately 20 nL of the solution on-column, shows the presence of two isomeric adducts generated by the cis/trans-opening of the epoxide ring upon reaction with deoxyguanosine (**Figure 1.20**). Analysis of the same solution by introduction of 2 μL of the same solution via sample stacking produced the electropherogram of **Figure 1.21** in which the two isomers are no longer resolved due to column overloading. However, despite the loss of resolution, the increased dynamic range revealed the presence of minor components, in this case identified as oligonucleotide adducts which were present because of incomplete digestion of the DNA.

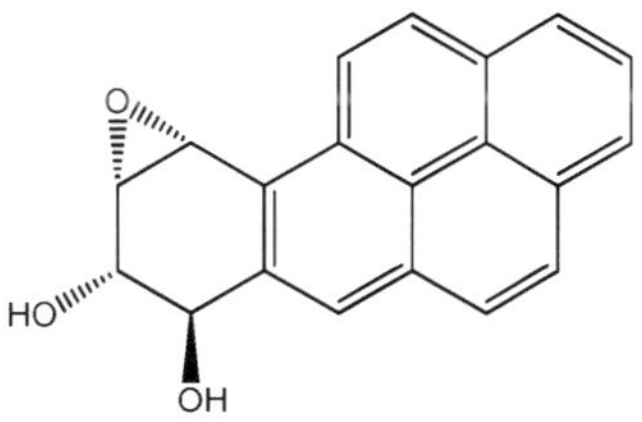

Figure 1.18. Structure of *anti*-7,8,9,10-tetrahydrobenzo[*a*]pyrene-7,8-diol 9,10-epoxide (BPDE).

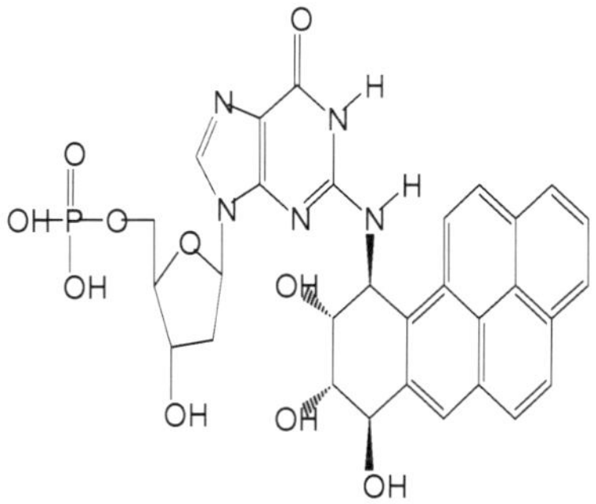

Figure 1.19. Structure of BaP adduct of deoxyguanosine monophosphate

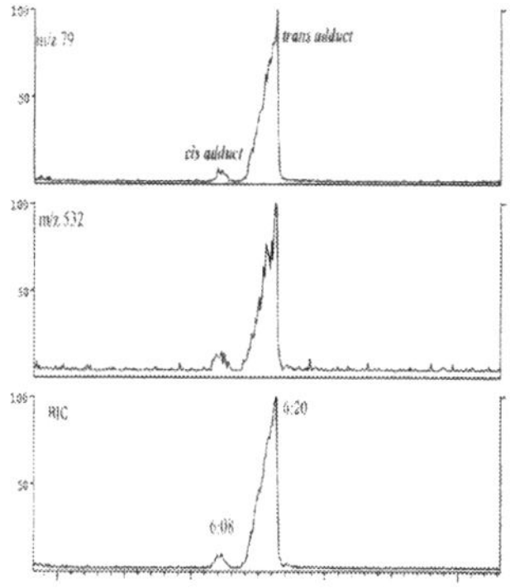

Figure 1.20. CZE electropherogram with SRM detection (*m/z* 648 → *m/z* 532) and (*m/z* 648 → *m/z* 79), showing two N^2-dGMP-BaP adducts (Barry et al.. 1996b)

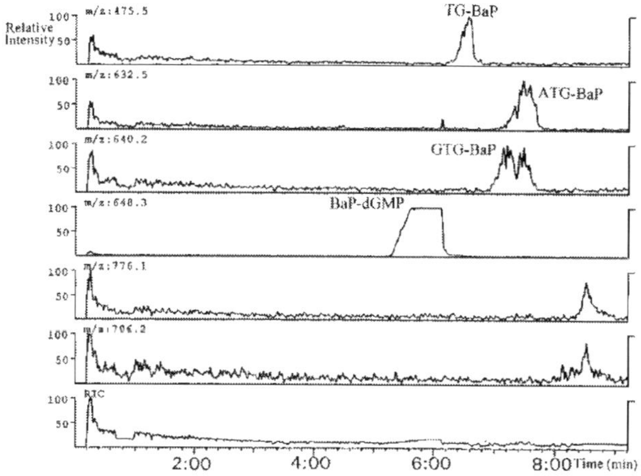

Figure 1.21. Mass electropherogram of in vitro reaction mixture. Mass spectrometer was scanned from m/z 400 to m/z 1400 (1 scan/2 s) (adapted from Barry et al., 1996b).

In another paper, Barry et al. (1996a) described the CE-MS analysis of a series of modified oligonucleotides four, five and six bases in length using a coated fused-silica column filled with an aqueous solution of polyvinylpyrrolidone (PVP). Oligonucleotides of the same length and sequence but differing only by the presence or absence of a small modification (such as a methyl group) on a single base were readily resolved. In a subsequent publication, Harsch and Vouros (1998) further improved upon this methodology to identify isomeric AAF-modified oligonucleotides as shown in **Figures 1.22** and **1.23**.

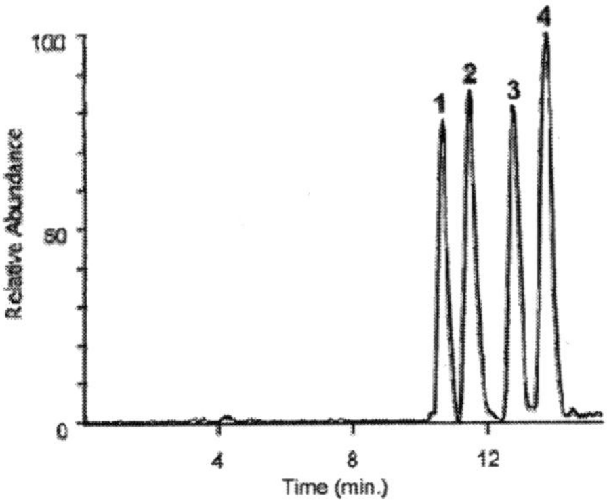

Figure 1.22. CE/MS in a PVP matrix. Analysis of four isomeric, AAF-modified oligonucleotides: ATG*CTA, ATTCAG*, TG*TAAC, CATG*AT. Field strength, ~450 V/cm. On-line MS detection of the doubly charged molecular ion at *m/z* 1005 (Harsch and Vouros, 1998).

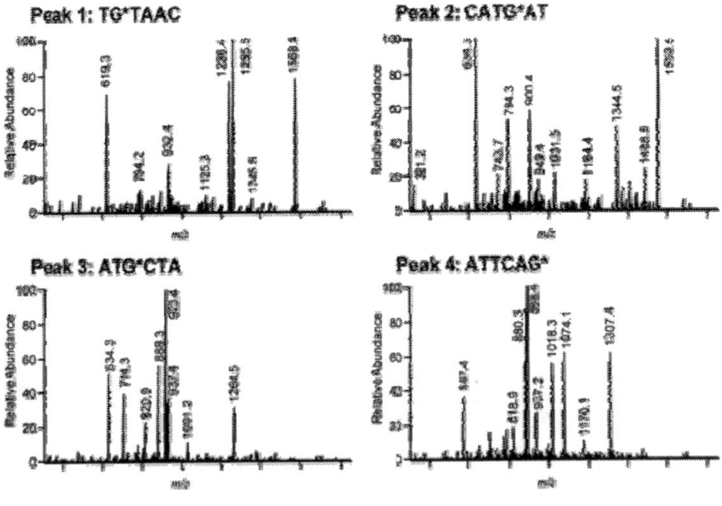

Figure 1.23. Acquired on-line MS/MS spectra (CID of *m/z* 1005) of four AAF-modified, isomeric oligonucleotides (Harsch and Vouros, 1998).

Chromatographic Separation Methods

Ion-Pairing Reversed Phase HPLC (IP-RP-HPLC)

Nucleotide analysis has proven to be cumbersome for RP-LC owing to the polar nature of the phosphate groups. Due to the acidity of the phosphodiester bridges, DNA dissociates into polyanions in aqueous or hydro-organic solution as well as in the gas phase. Although adducts are quite stable in neutral or slightly basic solution, loss of the nucleobases followed by backbone cleavage at acidic pH or hydrolysis of the phosphodiester groups at basic pH results in rapid degradation of DNA. Ion pairing reversed phase HPLC (IP-RP-HPLC) can overcome many of these problems by virtue of its high-resolving capability for DNA and the ability to use a variety of ESI-MS-compatible mobile phases.

The magnitude of electrostatic interaction, and thus retention, is determined by several factors, including: 1. hydrophobicity of the column packing. 2. charge. 3. hydrophobicity and concentration of the pairing ion. 4. ionic strength, temperature and dielectric constant of the mobile phase. 5. concentration of organic modifier. 6. charge and size of the DNA molecule. Elution of the adsorbed DNA is effected by a decrease in the surface potential due to desorption of the amphiphilic ions from the stationary phase with a gradient of increasing organic modifier concentration. Because the number of charges uniformly increases with size, double stranded (ds) DNA molecules are separated according to chain length in IP-RP-HPLC. The separations of DNA are usually more efficient at elevated temperature such as between 25 ^{o}C and 80 ^{o}C to reduce the probability of formation of double stranded oligomers.

In general, the criteria for HPLC analysis of oligonucleotide adducts have been based on those developed for separation of unmodified oligonucleotides. Challenges associated with the chromatographic analysis of oligonucleotides have been reviewed by Huber and Oberacher (2002). For example, Belicher and Bayer separated oligonucleotides up to 24-mers by RP-HPLC in a 100 x 2 mm i.d. Nucleosil C18 column with gradients of acetonitrile in 10 mM ammonium acetate (Bleicher and Bayer, 1994a, 1994b). However, more efficient separations can be accomplished via the use of various trialkyl- and tetraalkylammonium salts. The positively charged, hydrophobic triethylammonium ions are adsorbed onto the nonpolar surface of the stationary phase, resulting in the formation of an electric double layer that has an excess of positive charges near the surface.

Huber and Krajete (1999) used 100 mM triethylammonium acetate:acetate for the separation of single stranded (ss) oligonucleotides with single nucleotide resolution ranging in size from 3-mers to longer than 80-mers. Other reagents include diisopropylammonium acetate used to separate oligonucleotides up to 20-mers in a 150 x 0.5 mm i.d. column packed with a polymeric PLRP-S 5 μm 100 A stationary phase in acetonitrile-water (Bothner et al., 1995). Apffel et al. (1997a, 1997b) analyzed oligonucleotides up to 75-mers by using 400 mM 1,1,1,3,3,3-hexafluoro-2-propanol (HFIP)/2.2 mM TEA as ion-pairing reagents and methanol as organic modifier. Triethylammonium (TEAB) and butyldimethylammonium bicarbonate (BDAB) have provided attractive alternatives in several applications (Huber and Krajete, 1999; Oberacher et al., 2001a, 2001b).

Monolithic Columns

The technique's history and developments were reviewed recently by Svec and Huber (2006). Monolithic columns are noted for their extremely fast chromatographic separations at high flow rates and at reasonably low back pressure. The high throughput feature is very advantageous towards biomolecule analysis, as the mass transfer of large molecules such as polymers, peptides, proteins and nucleic acids is considerably slower than those of small molecules. This is because the diffusion coefficients of these high molecular weight compounds are several magnitudes smaller than those of low molecular weight compounds. In a direct comparison, a monolithic column made of polystyrenedivinyl benzene (PSDVB) phase performed 30%-40% better than 2 μm, nonporous PSDVB counterparts. Monolithic columns have unique properties such as tolerance to high flow rates, rapid speed of separation, and combined with nano LC can achieve very high sensitivity. A great deal of activity is currently focused on the optimization of monolithic columns for separation of small and large molecules.

Premstaller, Oberacher, and Huber used their own synthesized monolithic columns to achieve rapid and highly efficient separation of ss oligonucleotides and ds DNA fragments by IP-RP-HPLC (Premstaller et al., 2000). They obtained an improvement of approximately 40% in column performance, enabling the separation and detection of ds DNA fragments ranging in size from 51 to 587 base pairs by IP-RP-HPLC-ESI-MS (**Figure 1.24**). They also applied the online IP-RP-HPLC-MS/MS to sequence short oligonucleotides. They demonstrated that the chromatographic separation performance for monolithic capillary columns for ss and ds nucleic acids significantly surpasses that of capillary columns packed with microparticulate stationary phases.

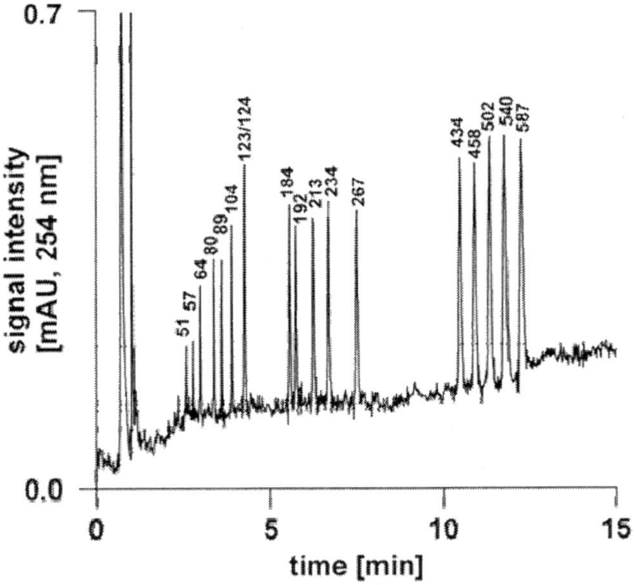

Figure 1.24. High-resolution capillary IP-RP-HPLC separation of a mixture of ds DNA fragments in a monolithic capillary column. Column, continuous PS-DVB, 60 × 0.20 mm i.d.; mobile phase, (A) 100 mM TEAA, pH 7.00, (B) 100 mM TEAA, pH 7.00, 20% acetonitrile; linear gradient, 35-75% B in 3.0 min, 75-95% B in 12.0 min; flow-rate, 2.2

uL/min; temperature, 50 °C; detection, UV, 254 nm; sample, pBR322 DNA-*Hae* III digest, 1.81 fmol of each fragment (adapted from Premstaller et al., 2000).

In a number of cases, researchers have shown that the new monolithic materials outperform traditional columns in terms of chromatographic resolution, separation speed, and other factors. Several monoliths based on organic and inorganic polymers have shown to be efficient in separating proteins, peptides, oligonucleotides, and other analytes. Scientists can optimize two sets of parameters simultaneously by using chemical methods to polymerize liquid precursors into a continuous porous mass of coalesced particles. One set are the parameters relating to the nature of material, porosity and other properties that affect the separation, another set are the size of the channels and open spaces that dictate the materials' permeability and give them their sponge-like structures. Effective monolithic columns could be prepared from styrene, divinylbenzene, and other monomers. Higher separation efficiency goes hand-in-hand with higher permeability and higher flow rates. It has been demonstrated that low-attomole and even zeptomole (10^{-21}*mol*) detection sensitivity can be achieved for peptide/protein analysis using a monolithic column in LC-ESI-MS (Yue et al., 2007).

<u>Selected HPLC-MS/MS Applications to the Analysis of Oligonucleotide Adducts</u>

In analyzing oligonucleotide adducts, two fundamentally different approaches are being pursued. The first, introduced by Tretyakova et al. (1996), is an indirect approach and relies on the use of stable isotope labeling of individual bases to analyze the distribution of DNA adducts by detecting the nucleoside adducts level produced by stepwise hydrolysis of the parent modified oligonucleotide (Koc et al., 1999; Wang et al., 2001). The second involves the direct analysis of modified oligonucleotides using collision induced dissociation (CID) in tandem mass spectrometry. We will discuss the two methods with primary emphasis on the second approach since it is of specific relevance to the research in this work.

Indirect Determination of Adduction Site in Modified Oligonucleotides

Much of the work of Tretyakova and co-workers at the University of Minnesota Cancer Center has focused on adducts of tobacco specific carcinogens (Lao et al., 2005, 2006a, 2006b; Rajesh et al., 2005; Sturla et al., 2005; Upadhaya et al., 2003; Wang et al., 2003; Ziegel et al., 2003, 2004). The roles of DNA adducts of NNK, NNAL, BaP, BPDE, POB in carcinogenesis with relation to sequence context at specific guanines within *K-ras* and *p53* gene sequences which are the mutation hotspots in lung tumors of smokers were investigated.

Tretyakova et al. (2002) described in detail the stable isotope labeling HPLC-ESI-MS/MS approach they developed which is universally applicable to studies of modifications to isolated DNA by other carcinogens and alkylating agents. This method simultaneously provides information on the chemical identities and sequence context of carcinogen-induced nucleobase adducts. They observed that N^2-BPDE-dG adduct formation is strongly affected by local sequence environment in the *K-ras* gene and there are important differences between the distributions of lung-cancer-associated mutation and the N^2-BPDE-dG adduct within a *p53* exon 5 derived gene.

First, ^{15}N-labeled guanine nucleobases were placed at specific positions within DNA oligodeoxynucleotides representing gene sequences of interest. Then they treated the oligonucleotides and DNA with adducts, and subjected them to enzymatic digestion to break down to mononucleoside adducts. These adducts were subsequently analyzed by HPLC-ESI/MS/MS in the positive ionization mode. The adducts formed at the ^{15}N-labled guanine could be distinguished from the other adductions sites by their molecular weight, which was increased due to the presence of the ^{15}N atom. The quantitation of the nucleoside adducts was done by MRM, based on the facile loss of deoxyribose (M=116) under the CID conditions. As demonstrated in **Figure 1.25**, the extent of adduct formation at the isotopically labeled nucleobase was calculated directly from the areas under the SRM peaks corresponding to ^{15}N and unlabeled adducts, respectively.

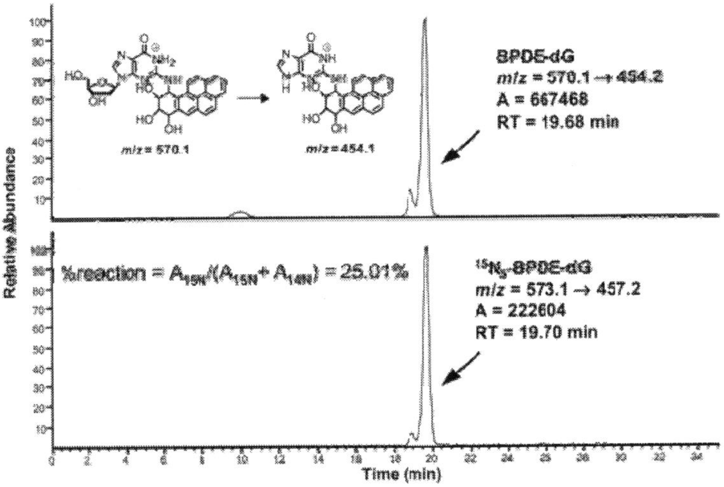

Figure 1.25. HPLC-ESI-MS/MS analysis of N^2-BPDE-dG in [$^{15}N_3$]-labeled *K-ras* derived DNA fragment, CCC GGC ACC ^{Me}CXC GTC CGC G (X = [$^{15}N_3$-dG]). Selected reaction monitoring was performed using the transitions: BPDE-dG, *m/z* 570.1 →*m/z* 454.0; [$^{15}N_3$]-BPDE-dG, *m/z* 573.1 →*m/z* 457.0. HPLC: Agilent 1100 series capillary liquid chromatograph (Agilent technologies). A Zorbax SB-C18 column (150 × 0.5 mm, 5 μm, Agilent Technologies) was eluted at a flow rate of 15 μL/min. HPLC solvents: A = 33% methanol in 15 mM ammonium acetate, B = acetonitrile, gradient 0-30% B in 22.5 min. MS: Finnigan MAT TSQ 7000 (ThermoQuest, San Jose, CA) operated in ESI$^+$ mode. Spray voltage, 5 kV; collision gas pressure, 2 mT; heated capillary, 200 °C; electron multiplier, 2200 V (adapted from Tretyakova et al., 2002).

In this study, the authors compared the reactivity of different guanines within *p53* exon 5 toward BPDE by preparing synthetic oligodeoxynucleotides CC$\mathbf{G_1}$$G_2$CACCC$\mathbf{G_3}C\mathbf{G_4}TCC\mathbf{G_5}CG_6$ containing [^{15}N] label at one of the highlighted positions, $\mathbf{G_1}$, $\mathbf{G_3}$, $\mathbf{G_4}$, or $\mathbf{G_5}$ (codon 157 = G_4TC, codon 158 = CG_5C). They then replaced cytosine nucleobases with 5-Me-C in accordance with the patterns of endogenous cytosine methylation within CpG dinucleotides of the *p53* gene. They observed higher-than-average reactivity toward BPDE for all four methylated positions $\mathbf{G_1}$, $\mathbf{G_3}$, $\mathbf{G_4}$, and $\mathbf{G_1}$. The relative abundances of N^2-BPDE-dG adducts were in the following order: **G3** (codon 156) >> **G5** (codon 158) > $\mathbf{G_1}$ (codon 154) ≈ $\mathbf{G_4}$ (codon 157) (**Figure 1.26**).

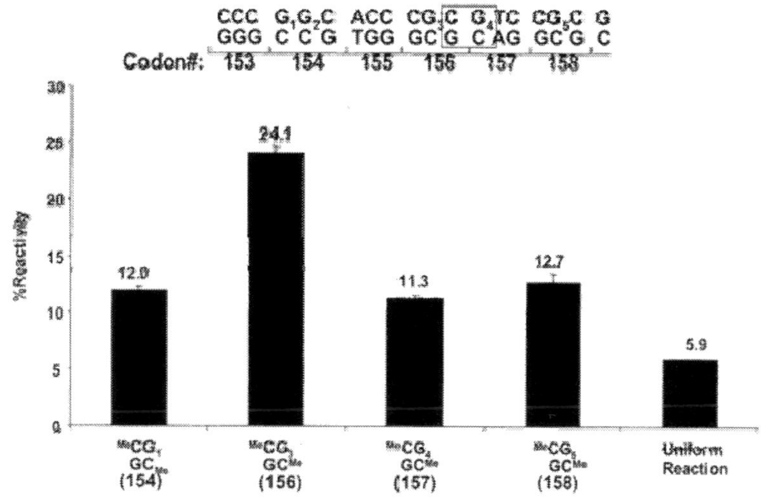

Figure 1.26. Relative formation of N^2-BPDE-dG at guanine nucleobases within a double-stranded oligodeoxynucleotide representing a region *p53* exon 5 containing frequently mutated codons 157 and 158*:* $CCG_1G_2CACCCG_3CG_4TCCG_5CG$). The data were compiled from two separate experiments (N = 3-6). The relative adduct formation at each guanine was calculated from the area ratio of HPLC-ESI-MS/MS peak corresponding to [^{15}N]-N^2-BPDE-dG to the sum of unlabeled and [^{15}N]-adduct peak areas. The random reaction value was determined from the total number of guanine nucleobases in both DNA strands (adapted from Tretyakova et al., 2002).

This paper was their first report to fully describe a mass spectrometry based approach that can directly quantify the formation of DNA adducts at specific positions within DNA sequences to investigate the sequence selectivity of the adducts. Their study, however, revealed a complex relationship between DNA sequence and reactivity toward adducts which cannot be simply explained by the neighboring nucleobases. Most recently Guza et al. (2006) applied LC-ESI-MS/MS to study the kinetics of O6-methyl-dG repair by O6-alkylguanine DNA alkyltransferase within *K-ras* gene-derived DNA sequences.

Direct Characterization of Oligonucleotide Adducts by MS/MS and LC-MS/MS

The method of Tretyakova and coworkers represented a dramatic improvement in both sensitivity and adduct quantitation over traditional bioanalytical methods such as PAGE and fluorescence. However, the extensive sample preparation and handling in addition to digestion of oligonucleotides to monomers suggests that a direct method based on characterization of the intact oligonucleotide adduct by collision induced dissociation (CID) would be more efficient for determination of the carcinogen binding spectrum within the nucleotide sequence. Tandem MS using CID was the obvious route and, in accordance with this reasoning, Iannitti et al. (1997) at the University of Wollongong of Australia used ESI-MS and ESI-MS/MS to examine the sequence selectivity of adducts formation between the antibiotic hedamycin and model 6-mer oligonucleotides. Samples were purified off-line and introduced by direct infusion into the ESI-MS/MS system. In a follow-up paper, Colgrave et al. (2003) determined the sequence selectivity and stability of alkylated oligonucleotide adducts of duocarmycin C_2 (pyrindamuycin A), duocarmycin C_1 (pyrindamuycin B) hedamycin, and DC92-B by ESI-MS and ESI-MS/MS offline. Glover et al. (1998) used tandem MS to fully sequence a 7-mer para-benzoquinone adducted oligonucleotide. They compared the different fragmentation pathways on different charge states and CID collision energy. Their results demonstrated the modified oligonucleotides follow the same fragmentation patterns of unmodified oligonucleotides as discussed in the previous section.

Concurrently, in our laboratory, Marzilli et al. (1998) used ESI-ITMS (electrospray ionization ion trap mass spectrometry) to unambiguously identify three isomeric oligonucleotides of the sequence 5'-CCGGAGGCC modified by carcinogen aflatoxin B_1 and the site of the adduction by MS, MS/MS and MS^3. In **Figure 1.27**, product ions of the $[M-3H]^{3-}$ ion at *m/z* 1016 revealed the site of base modification to be the seventh nucleotide from the 5′ end of the oligonucleotide.

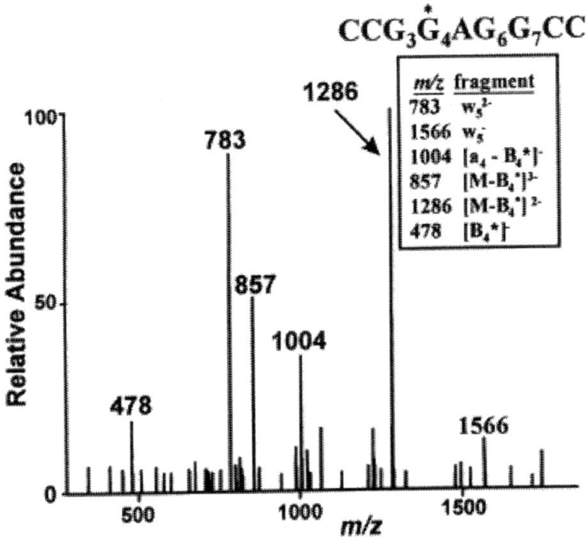

Figure 1.27. MS/MS spectrum of the $[M-3H]^{3-}$ ion at *m/z* 1016 of AFB_1–5′–CCGG*AGGCC adduct peak C. The modified base is designated with an asterisk (Marzilli et al., 1998).

Furthermore, by employing MS^3 to the selected $[M-B_4*]^-$ ion at m/z 1286, two complementary product ions were generated (**Figure 1.28**). This collision spectrum provided further evidence that the AFB_1 modification was on the G4 position along the oligonucleotide chain.

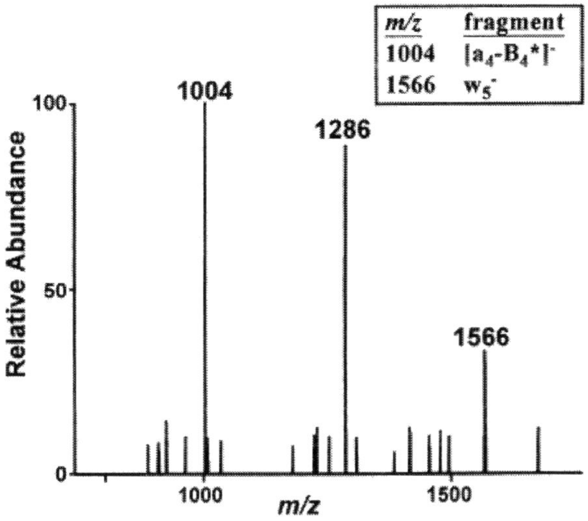

Figure 1.28. MS^3 spectrum (*m/z* 1016 → *m/z* 1286 → ?) of $[M\text{–}B^*_4]^{2-}$ ion. The result is cleavage of the phosphodiester backbone at the site of modification (Marzilli et al., 1998).

The results reported above demonstrated the utility of CID-MS/MS for mapping adduct distribution in short oligonucleotide sequences. However, all these investigations were conducted using model oligonucleotide sequences purified off-line and introduced into the MS by direct infusion. A natural follow-up to that was the coupling of CID with chromatographic separation and the challenging separation of the resulting positional isomers. As a first step toward that goal, a number of studies have been carried out in which short oligonucleotides (generally <15-mers) were reacted either as single- or double-stranded with reactive metabolites of carcinogens and the reaction products analyzed by HPLC-MS/MS to determine the sequence selectivity of DNA alkylating agents.

In work from our laboratory, the ds 12-mer sequence $TAGTCA^{579}A^{580}GGGCA$ from the coding region of the *hprt* gene in Chinese hamster V79 cells, was reacted with benzo[a]pyrene diol epoxide (BPDE), the reactive metabolite of the carcinogen benzo[a]pyrene (B[a]P) (Harsch et al., 2000). In this sequence the two indicated dA nucleotides are known to differ markedly in their extents of mutation after treatment with the carcinogen epoxide. Analysis by RP-HPLC-ESI-MS/MS allowed assignment and semi quantitative analysis of the sites of modification for positional isomers. **Figure 1.29** displays the TIC derived from the RP-HPLC-ESI-MS/MS analysis of the 12-mer modified with BPDE. The sites of adduct formation in the separated compounds were identified by an analysis of the characteristic series of a-B and w fragment ions generated by CID of the triply deprotonated molecule. It is seen that dA adducts predominated over dG adducts in the case of ds DNA, whereas the reverse was true for ss adducts (**Figure 1.30**). Additionally, the method permitted a facile distinction of the two strands of an oligonucleotide duplex, based on the differing masses of the strands. In consequence, adduct formation in both strands of a DNA duplex with multiple target bases can be successfully detected by RP-HPLC-ESI-MS/MS. This study represented the first example of the use of LC-MS/MS for the identification of adduct-containing sequences with multiple target bases in a ds oligonucleotide.

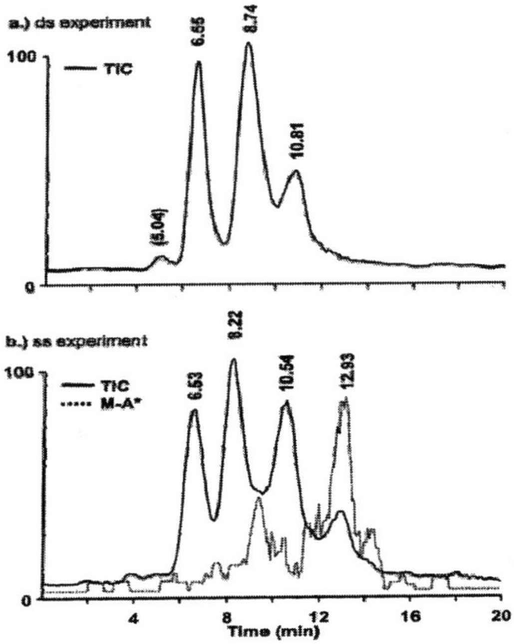

Figure 1.29. LC-MS/MS analysis of ds (a) and ss (b) samples. The TIC of the MS/MS fragmentation of BcPh DE-modified TAGTCAAGGGCA under CID conditions was monitored in both experiments. In addition, panel b displays the extracted mass data trace of M-A* corresponding to loss of the modified adenine (A*) from the full-length oligonucleotide by depurination under CID conditions (adapted from Harsch et al., 2000).

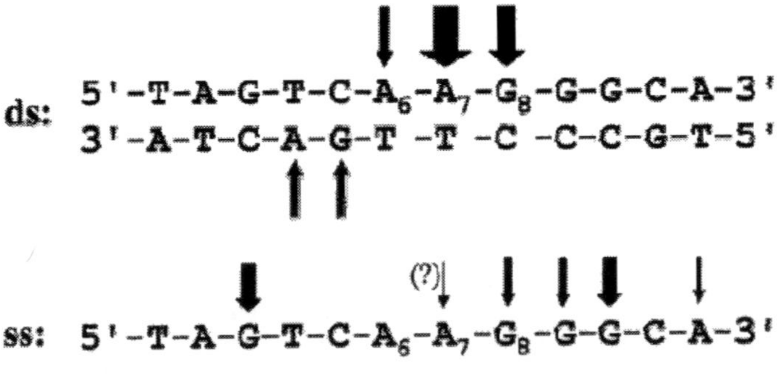

Figure 1.30. Identified sites of (-)-BcPh DE-2 adduct formation on the ds and ss oligonucleotides. Sizes of the arrows indicate an approximation of the relative amounts of adducts formed in each individual (ds or ss) sample, and do not imply a direct comparison between the two samples (adapted from Harsch et al., 2000).

In the Esmans group, Van den Driessche et al. (2004a) modified model oligonucleotides of 2 up to 8 bases in length and studied them with the aid of miniaturized LC coupled to ESI-MS/MS. They found that melphalan alkylation occurs in the sequence G > A > C > T. They chose a specific oligonucleotide sequence so that the difference in alkylation specificity could be observed (**Figures 1.31** and **1.32**).

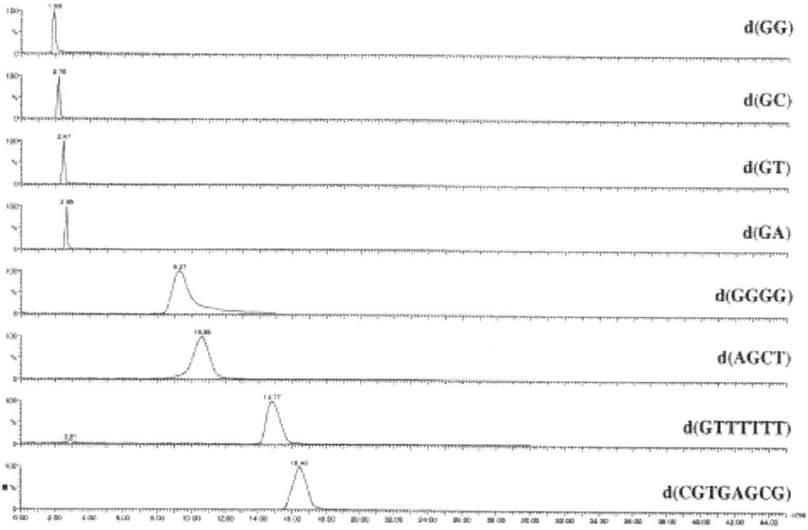

Figure 1.31. Chromatographic separation of a mixture of oligonucleotides on a RP column (C_{18}, 300 µm i.d., 5 cm length): d(GG), d(GC), d(GT), d(GA), d(GGGG), d(AGCT), d(GTTTTTT), and d(CGTGAGCG) (adapted from Van den Driessche et al., 2004a).

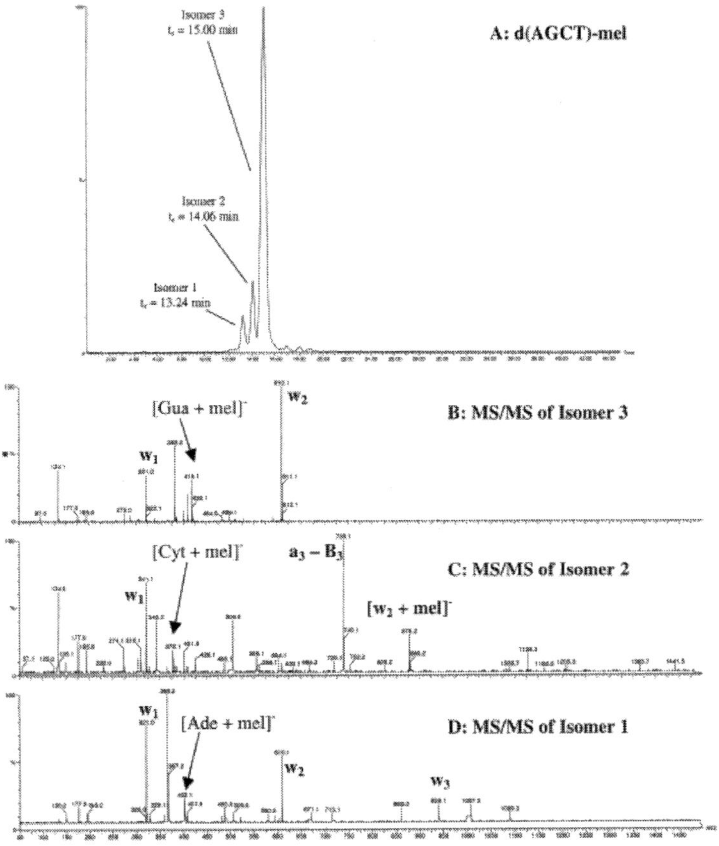

Figure 1.32. Trace A: Chromatographic separation of the reaction mixture of a melphalan alkylated tetra-mer d(AGCT). Isomers can be seen which represents alkylation on the adenine- (Trace D), cytosine- (Trace C), and guanine residue (Trace C). Alkylation on the thymine base could not be observed (adapted from Van den Driessche et al., 2004a).

In the Fu group of Guangzhou Research Center of Mass Spectrometry, Chen et al. analyzed DNA methylation by IP-RP-LC-MS/MS on model oligonucleotides of 6-mer, 8-mer and 10-mers. Ion-pairing LC achieved nice separation of unmodified and modified oligonucleotides and MS/MS confirmed the methylated sites (Chen et al., 2004a). Recently, Song et al. (2005) investigated the sites of modification by PGE on three synthetic ss oligonucleotides by IP-RP-LC-MS/MS, **Figure 1.33**. MS/MS was performed (**Figure 1.34**) to identify the oligonucleotide fragments via the characteristic series of fragment ions a-B. However, their study involved reaction of the carcinogen with ss oligonucleotides, which may not be physiologically representative as in a biological system the carcinogen is presented to the double stranded DNA.

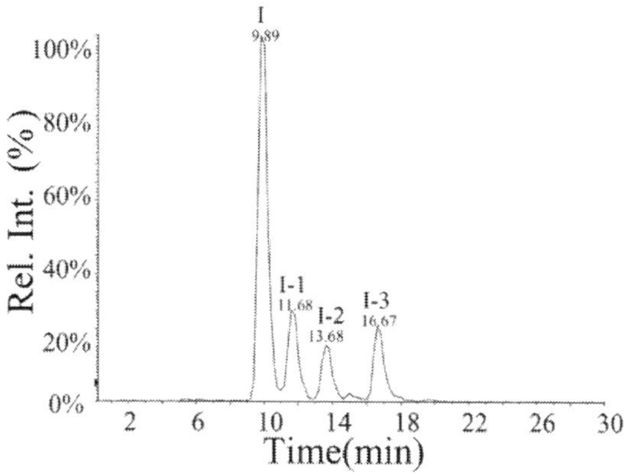

Figure 1.33. LC/negative ion ESI-MS chromatograms obtained from the injection of a reaction mixture of PGE with 5'-TATCTGTC-3'. Peak I is the unmodified oligonucleotide, while peaks I-1 to I-3 are PGE-modified versions (adapted from Song et al., 2005).

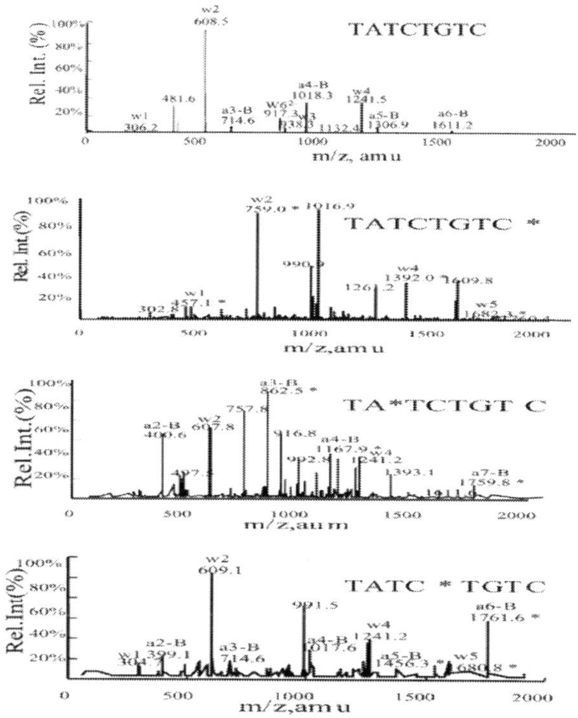

Figure 1.34. Fragment ion spectra for $[M-2H]^{2-}$ ions of unmodified I and oligonucleotides adducts I-1 to I-3, obtained by on-line HPLC/MS/MS. Asterisk * indicates bases carrying the modification (adapted from Song et al., 2005).

A significant improvement toward the use of HPLC-MS/MS for the analysis of oligonucleotide adducts has been realized via the use of monolithic columns as shown in a recent publication by Xiong et al. (2007), which employed a monolithic PSDVB column coupled to nanoESI-MS for separation and identification of isomeric oligonucleotide adducts derived from the covalent binding of BPDE to the ds 14-mer 5'-PO_4^--ACCCGCGTCCGCGC-3'/5'-GCGCGGGCGCGGGT-3'. The in-house fabricated monolithic column was able to effectively separate four positional isomeric BPDE oligonucleotide adducts which were identified subsequently by MS/MS. Moreover, based on the MS/MS fragmentation patterns it could be deduced that peaks III-1 and III-2 represent diastereomeric adducts in which (±)-BPDE is attached to the same guanine base (G_7) of the oligonucleotide. Peak III-3 represents the case in which (±)-BPDE adduct is attached to the guanine G_{11} of the oligonucleotides. Peaks III-4, III-6 and III-7 represent diastereomeric adducts in which BPDE is attached to the same guanine G_5 of the oligonucleotide. Peaks III-5, III-8 and III-9 represent diastereomeric adducts in which BPDE is attached to the same adenine A_1 of the oligonucleotide. Surprisingly, no adduct was found in which (±)-BPDE is attached to the guanine G_{13} of the oligonucleotides. The chromatogram depicted in **Figure 1.35** reveals the partial separation of at least nine peaks, with shoulders on some of the peaks indicating the presence of even more compounds. However, no further information could be obtained on the nature of the stereoselectivity of the adducts studied in this work.

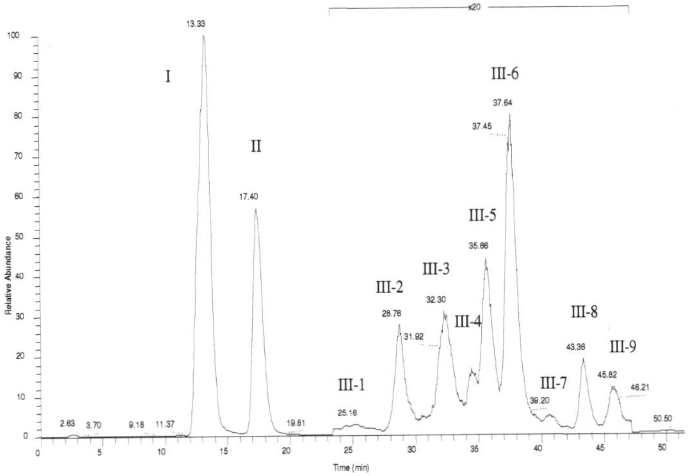

Figure 1.35. LC-MS composite EICs for the primary oligonucleotide strand obtained from the injection of a crude reaction mixture of (±)-BPDE with ds-oligonucleotide on an optimized monolithic PS-DVB nanocolumn (I) the unmodified 5'- phosphorylated oligonucleotide (m/z 1420) (II) the unmodified 5'-dephosphorylated oligonucleotide (m/z 1394) and (III) the modified 5'-phosphorylated oligonucleotides (m/z 1521) (adapted from Xiong et al., 2007).

HPLC-MS Analysis of Oligonucleotide Adducts Produced from Enzymatic Digestion of Modified Oligomers

It is evident from the above discussion that that HPLC-MS/MS represents an ideal method for investigating the chemical selectivity in DNA sequences that contain multiple target bases competing for adduction. However, most current studies have focused on the examination of model systems in which an oligomer of known sequence has been analyzed by HPLC-MS/MS following its reaction with an activated carcinogen. Of more relevance is its application to the analysis of in vitro or in vivo systems where the enzymatic digestion of DNA produces a mixture rich in fragments containing the covalently bound carcinogen. Here the challenges are not only the selection of the proper enzymatic digestion process but also the complexity of the mixtures which require use of high resolution chromatography and the ability to handle the voluminous amount of data from their LC-MS analysis. It is with these factors in mind that the current study was undertaken in order to advance the investigation of the sequence context of carcinogenic binding to DNA from simple model studies and bring it closer to in vitro and ultimately in vivo applications.

The significance of the direct LC-MS/MS approach for the analysis of oligonucleotide adducts produced from the enzymatic digestion of DNA or large oligomers to map adduction sites, was first described in a 1994 publication by Janning et al. (1994). Modified DNA was digested to short DNA fragments ranging in length from dinucleotides to hexanucleotides using benzonase/alkaline phosphatase and styrene oxide adducts were characterized by CZE-MS/MS in negative ion mode. The dinucleotides were present as singly charged species while trinucleotides were detected as doubly and triply charged species. In subsequent publications, the Linscheid group determined styrene oxide (SO) adducts in DNA in the form of oligonucleotides at lengths of 2-8 bases by CZE-MS and LC-ESI-MS (Schrader and Linscheid, 1995, 1997). Wang et al. (1999) used Nuclease P1 to digest oligonucleotide photoproducts to small photoproduct-containing trinucleotides whose structures were assigned on the bases of molecular weight and MS/MS data. The partial sequence information of these oligo fragments was preserved by such a method and this approach paved way to do nearest-neighbor analysis and identify hotspots for photodamage in cellular DNA (**Figure 1.36**). The authors suggested the method to be generally useful for studying all types of DNA modifications. More recently Chou, et al identified a series of 6,7-Dihydro-7-hydroxy-1-hydroxymethyl-5H-pyrrolizine (DHP) modified dinucleotide adducts of riddelline by negative ion LC-ESI-MS/MS. Adducted calf thymus DNA was enzymatically digested by micrococcal nuclease (MN) and spleen phosphodiesterase (SPD) and separated by HPLC (Chou et al., 2003). In **Figure 1.36**, the left panel shows an example of the MRM chromatograms for the set of isomeric adducts derived from DHP modification of either adenine or thymine base in ApTp and TpAp in intact ctDNA. The right panel shows the corresponding mass chromatograms for product ion scans from the reaction of DHP with either ApTp and TpAp.

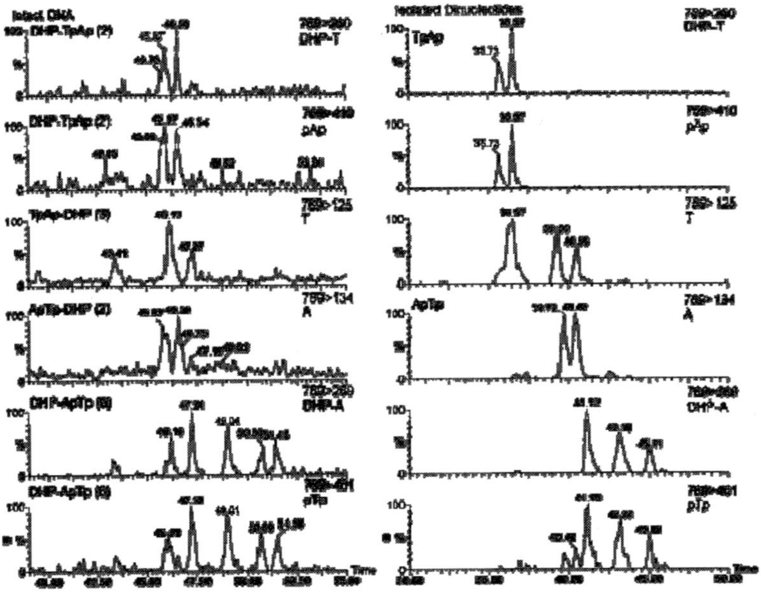

Figure 1.36. Characterization of representative DHP-modified dinucleotides from either isolated dinucleotides or intact ctDNA. Product ion chromatograms of selected transitions that characterize DHP adducts of TpAp and ApTp produced by reaction of DHP with the isolated dinucleotides (either TpAp or ApTp, right panel) or MRM chromatograms of the same adducts derived from DHP modification of intact ctDNA (left panel). The transitions monitored and the associated structural assignments are listed at the right of each chromatogram (Chou et al., 2003).

In 2004, Andrews et al. (2004) analyzed the in vitro digestion products of oligonucleotides and DNA modified by AAAF using IP-RP-LC-MS and LC-MS/MS. Benzonase/alkaline phosphatase was used to digest the oligonucleotides and DNA into fragments of different size, and allowed examination of the sequence context of adducted segments to determine binding preference of adduct to DNA by presenting the entire molecule as a target. The methodology produced primarily trimer fragments, the profiles of several of which are shown in **Figure 1.37**. Comparisons of MS/MS spectra with reference standards were made to confirm sequence determination and peak assignment from characteristic fragmentation patterns. The results pointed to the possibilities of an uneven base distribution in the DNA sequence.

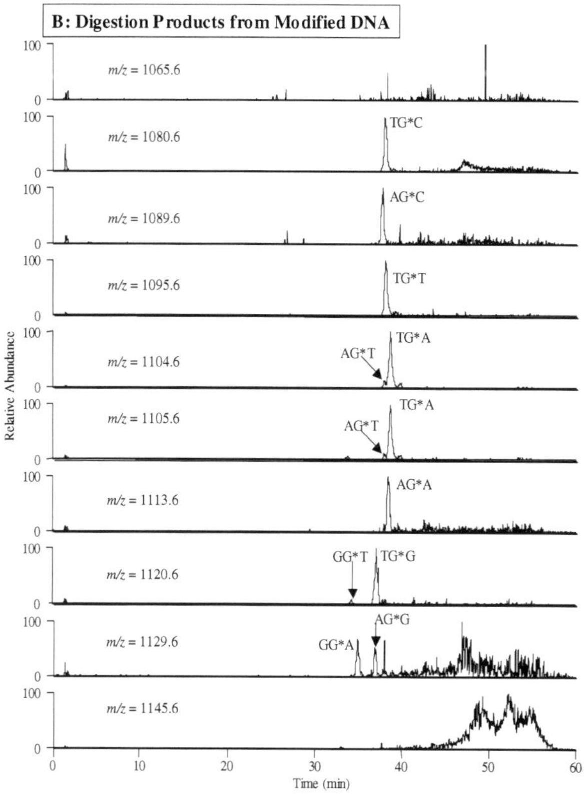

Figure 1.37. LC/MS chromatograms of modified trimers produced from the digestion of AAF-modified DNA (Andrews et al., 2004).

In summary, the assignment of the sites of modification within a target sequence is very important due to the fact that isomeric DNA adducts may possess different mutagenic activities. Knowledge of the sequence selectivity of adduct formation and the mutational consequences of such adduct formation is of paramount importance that can help us: i. develop biomarkers of effect associated with the carcinogen exposure, ii. identify the nature of the most mutagenically potent constituents of the complex mixtures of carcinogens to which humans are exposed and, iii. determine the risk associated with the carcinogen exposures. Although there have been a lot efforts, a sound approach capable of detecting unknown DNA adducts, preserving the base sequence information, and even the supramolecular requirements of the modification reaction still remains to be developed.

Although prior studies, including those from our own laboratory, have been successful in generating informative data on short (trimer) adducts, the acquisition of more definitive data, especially on longer oligonucleotide adducts was hampered by the inability to handle manually the information generated from the LC-MS/MS analysis of the digest. Therefore, the research undertaken in this thesis had as an objective the development of methodology which will ultimately lead to a more comprehensive approach to the characterization of oligonucleotide adducts produced from the enzymatic digestion of DNA. As a first step toward that goal it was determined that two key requirements had to be considered before we could address in a comprehensive fashion the susceptibility of a gene for adduction and its predilection for mutation.

1. The development of DNA enzymatic digestion procedures that would cut DNA into longer fragments, providing more complete information about neighboring base effects on adduction.

2. The development of data handling software to handle the myriad of data produced from the LC-MS analysis of DNA digests, bearing in mind that the generation of progressively longer oligonucleotides is likely to yield a multitude of isomeric species.

3. Investigate the utility of these procedures to evaluate the relationship between chemoselectivity and mutation in model oligonucleotides bearing a mutation hotspot to a specific carcinogen.

The results of these studies are described in the following chapters.

Chapter 2

Investigation of Enzymatic Behavior of Benzonase/Alkaline Phosphatase in the Digestion of Oligonucleotides and DNA by ESI-LC/MS

2.1 Introduction

One aspect of cancer research involves the study of DNA and DNA-adducts. The techniques in studying DNA include restriction analysis, polymerase chain reaction (PCR), molecular cloning, DNA sequencing, in situ hybridization and DNA chips, among others. Along these lines, it is important to develop effective techniques for the isolation, separation, quantitation, and structural analysis of DNA on par with the rapid rises in the field of genomics science. In DNA-adducts analysis, commonly used techniques are ^{32}P-postlabeling, immunoassays, fluorescence spectroscopy and mass spectrometry (Andrews et al., 1999; Esaka et al., 2003; Garner, 1998; Stiborova et al., 2004; Turesky and Vouros, 2004). Due to its improved sensitivity, mass spectrometry has been gaining more ground in DNA and DNA-adducts analysis in recent years (Guo, 1999; Soglia et al., 2001; Tost and Gut, 2006; Wolf and Vouros, 1994). The developments in mass spectrometry instrumentation have also increased its mass accuracy dramatically. In addition, mass spectrometry provides sufficient specificity and ability for elucidation of modified DNA structures. When coupled to separation techniques such as HPLC and CE, the MS technique is particularly advantageous in analyzing oligonucleotides and DNA adducts in complex mixtures (Andrews et al., 2004; Hoes et al., 1999; Soglia et al., 2001; Van den Driessche et al., 2003; Wolf and Vouros, 1994). The procedure often starts with enzymatic digestion (typically DNase I or Nuclease P1) to deoxynucleosides or deoxynucleotides prior to further analysis by MS (Gangl et al., 2001; Hoes et al., 1999; Wang et al., 1999; Wolf and Vouros, 1994). However, DNA sequence information is lost as the oligomers are cleaved to individual deoxynucleosides or deoxynucleotides. It has been observed that reaction of carcinogens with DNA often exhibits sequence selectivity, the recognition of which may be important in order to understand their mutagenic activities (Beleguise-Valladier and Fuchs, 1991; Hartley et al., 1992; Koehl et al., 1989a; Kohn et al., 1987; Mattes et al., 1986; Van den Driessche et al., 2004b). Hence it is advantageous to be able to obtain the sequence information from the analysis. The ability of electrospray (ESI)-MS to form multiply charged ions in combination with improved mass range in modern instruments, has made possible the detection and structural elucidation of larger digested fragments. The significance of this LC-MS approach was first described in a 1994 publication by Janning et al. (1994). However, there has been little follow-up on this topic and this may be partially due to the lack of suitable enzymes for digesting oligonucleotides, DNA and their adducts into sequence informative oligomeric fragments of size suitable for routine LC-MS analysis (Andrews et al., 1999: Wang et al., 1999).

In general, there are two types of enzymes, specific and non-specific. The specific enzymes work by recognition of a specific DNA sequence. However, while working with an unknown sequence, a non-specific, random endonuclease is preferred. There are a number of candidates in this category, whose enzymatic behavior is not fully understood. Among these, benzonase has been studied to some extent in the past (Meiss et al., 1995; Nestle and Roberts, 1969; Schrader and Linscheid, 1995, 1997). This enzyme is an extracellular endonuclease genetically engineered from the pathogenic Gram-negative bacteria *Serratia marcescens.* It is produced and purified from *E. coli* strain W3110, a mutant of strain K12, as a precursor of 266 amino acids that is proteolytically processed upon secretion to yield a non-covalent dimer of 245 amino acids. Each subunit is ~ 30 kDa and requires two disulfide bonds for its activity and is effective over a wide range of operating conditions. Interestingly, benzonase is the only known endonuclease which degrades all types of nucleic acids. It is able to cleave single stranded, double stranded, linear and circular DNA and RNA, producing 5'-monophosphate terminated di-, tri-, tetra-, and penta-nucleotides, along with a very small fraction (less than 2%) of mononucleotides. It is considered a nuclease with broad substrate specificity. Although the enzyme is capable of cleaving at nearly all positions along a nucleic acid chain, it was also reported that the nuclease does have sequence-dependent preferred cleavage sites (Meiss et al., 1995). It is known that the enzyme actually prefers GC-rich regions in DNA more than d(A)/d(T)-tracts. The optimum activity of this enzyme is achieved between 37 °C and 44 °C and is more non-specific at higher temperature and higher ionic strength.

In the study of oligonucleotides and DNA adducts the structural information of the digestion products can help to not only characterize the adducted sites but also provide information about the mechanism of enzyme-substrate recognition and reaction. Importantly, the enzymatic digestion procedure is non-destructive with respect to the adducted sites. Thus, a better understanding of the enzyme reaction mechanism will facilitate the selection of suitable enzymes in the study of oligonucleotides, DNA, and their adducts. It was found that the benzonase/alkaline phosphatase pair is non-specific and can digest oligonucleotides and DNA into different fragments varying in length from 2-mers up to 10-mers thus making it appropriate for the study of oligonucleotides and oligonucleotide adducts produced from DNA. Use of LC-MS revealed significantly more information than previously possible by other methods demonstrating that mass spectrometry can have an impact on understanding enzymatic reactions and their mechanisms.

2.2 Experimental Section

2.2.1 Chemicals

The single stranded (ss) oligonucleotide standards were obtained from Sigma Genosys (Woodlands, TX). The double stranded (ds) oligonucleotide standards were obtained from Integrated DNA Technologies, Inc. (Coraville, IA) and were desalted and purified by PAGE. Both 5' and 3' terminals ended with an OH group. The calf thymus DNA was purchased from Sigma-Aldrich (St. Louis, MO). Benzonase was purchased from Novagen (Madison, WI). Alkaline phosphatase was purchased from Sigma-Aldrich (St. Louis, MO). Citrate buffer (pH 6) was obtained from Sigma-Aldrich (St. Louis, MO). Tris/$MgCl_2$ buffer was obtained from Schwarz/Mann Biotech (Cleveland, OH). Triethylammonium bicarbonate (TEAB) was obtained from Sigma-Aldrich (St. Louis, MO). All solvents were HPLC grade unless specified otherwise. Acetonitrile, methanol, and ethyl acetate were obtained from Mallinckrodt Baker (Paris, Kentucky). Water was from EMD Chemical (Gibbstown, NJ). Oasis HLB 1cc (30 mg) SPE cartridges were obtained from Waters (Milford, MA).

2.2.2 Methodology

Sample Preparation

Stock Solution Preparation

The tubes containing the 4-mer standards or single stranded and double stranded oligonucleotide standards were centrifuged prior to opening. Then a volume of deionized (DI) water was added to the 4-mer oligonucleotide standards and the ss 12-mer 1, 12-mer 2, 20-mer, 21-mer, and 80-mer oligonucleotides to create 1 nmol/μL stock solutions. The ds 12-mer and 21-mer were resuspended with RNase-Free Duplex Buffer to make a 100 μM solution. The mixture was heated to 94 °C for 2 minutes, removed from heat and allowed to cool slowly to room temperature. This resulted in stable, double stranded form and was stored at -20 °C. One mg calf thymus DNA was weighed out and 1 mL 10 mM citrate buffer (pH 6) was added to produce a concentration of 0.1 nmol/μL. All samples were then vortexed.

Enzymatic Digestion

The oligomers and DNA were then dried down using a Speedvac from Savant (Pleasanton, CA) for 2 h at medium heat. They were reconstituted in 200 μL of 5 mM Tris/10 mM $MgCl_2$ buffer (pH 8). After being vortexed down, 5 μL benzonase (32.9 units) and 5 μL alkaline phosphatase (1.09 units) were added to the oligomers and DNA in buffer solution. The mixtures were vortexed and incubated for 24 h at 37 °C.

Chromatographic Conditions

The LC system employed was an Agilent Technologies 1090 HPLC equipped with an autosampler (Wilmington, DE). The HPLC system was controlled by HP ChemStation software, version A.06. The reverse phase RP-1 PSDVB (polystyrene divinylbenzene) column (2.1 mm x 100 cm x 3.5 μM) was purchased from Hamilton Chromatography (Reno, NV). The column was kept at ambient temperature (25 °C) during analysis. The mobile phase consisted of 25 mM TEAB as A and 25 mM TEAB with 20% acetonitrile as B. The gradient went from 0% B to 100% B in 20 minutes and was held at 100% B for 5 minutes. The flowrate was 0.2 mL/min and the injection volume was 20 μL.

Mass Spectrometric Conditions

MS data were acquired on a ThermoFinnigan LCQ Classic mass spectrometer (San Jose, CA). MS/MS data were acquired on a Waters Q-Tof micro mass spectrometer (Milford, MA). The mass spectrometers were tuned on a series of standard oligonucleotides. The samples were analyzed by negative ion electrospray mass spectrometry, in full scan mode in MS, and subsequently by MS/MS. The MS conditions on the LCQ were as follows: spray voltage 3.8 kilovolts, capillary voltage - 4 volts, capillary temperature 200 °C, N_2 sheath gas 25 arbitrary units, and tube lens voltage - 25 volts. The MS/MS conditions on the Q-Tof micro were as the follows: sample cone - 30 volts, capillary voltage - 2700 volts, collision gas argon, collision energy: 7 volts. MS data were acquired and processed with Navigator 1.2 software on the LCQ. MS/MS Data were acquired and processed with MassLynx 4.1 software on the Q-Tof micro. Mass spectra were recorded in continuum mode.

Computer Software

LC-MS data were analyzed using a computer software program specially designed for this project, and referred to here as GenoMass. The full details of the software will be reported in a forthcoming publication (Liao et al., 2008). The program automatically calculates masses of oligomers and/or their adduct-ions, searches for them through the LC-MS data, and then outputs a list of found fragments and their abundance and the corresponding base sequence along with other pertinent information. The program can also generate extracted ion chromatograms (EIC) based on the found masses on a click of the mouse. The software is designed to eliminate the otherwise extraneous manual labor demanded by the large volume of data produced by MS analysis of DNA and DNA-adduct mixtures. Here we give a brief explanation on how this custom-made GenoMass program used a "reversed pseudo-combinatorial" approach to analyze the LC-MS data. **Figure 2.1** shows the two input interfaces of the software when used to search in the LC-MS data of an unknown calf thymus DNA sample for all possible tetramers that were generated by the combination of the four bases A, T, G, C.

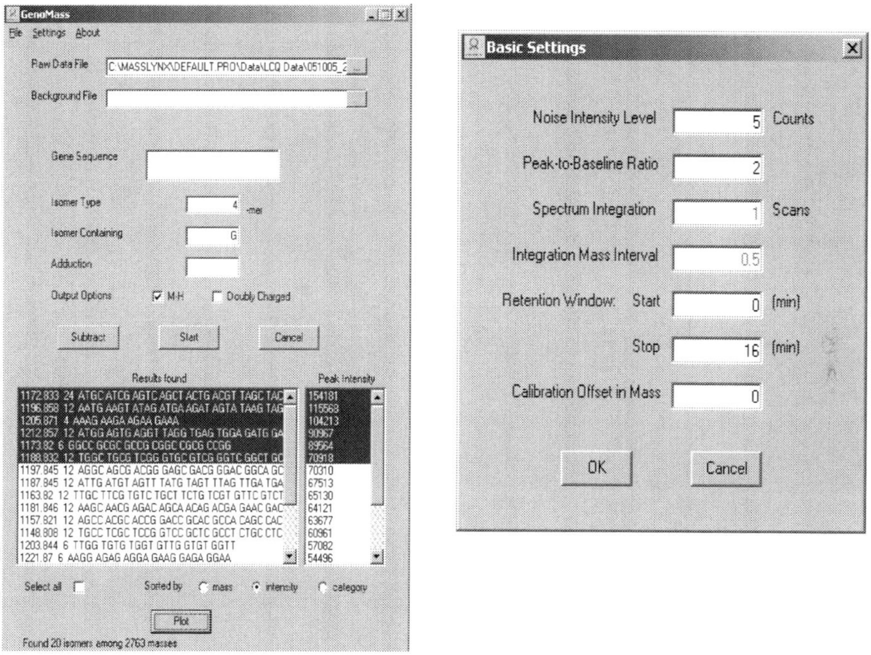

Figure 2.1. The interfaces used by GenoMass to search for all possible tetramer fragments in a calf thymus DNA digested by benzonase/alkaline phosphatase.

The program then listed found ion peaks by the order of intensity or mass within the defined retention window, and plotted the EICs of the selected ions. In this example, the six most intense ions corresponding to the sequences listed on the right of the masses were plotted automatically.

In the study of comparing orderly cut fragments to random cut fragments of enzymatic digestion products, GenoMass was able to generate a table of orderly cut fragments and random cut fragments with their corresponding mass, along with the sequences of all possible isomers and their occurrences. **Figure 2.2** shows the two interfaces of this operation.

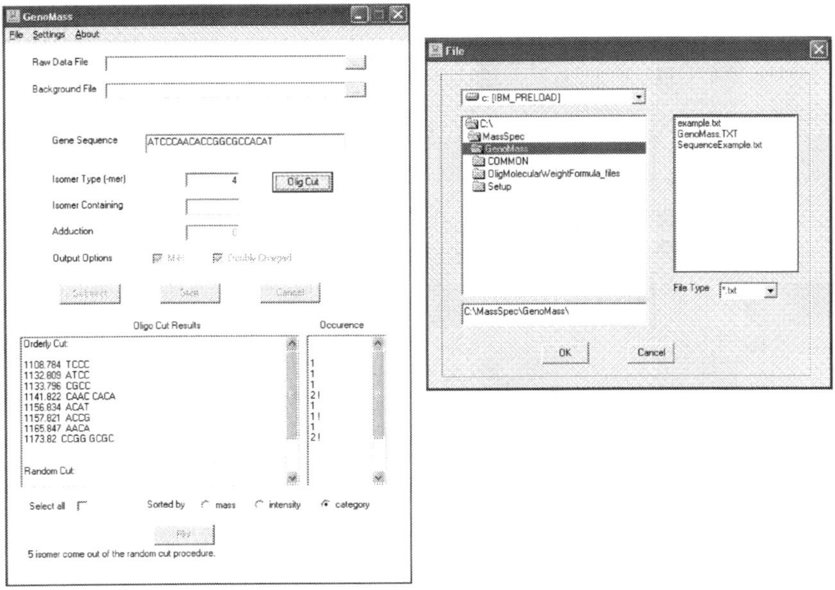

Figure 2.2. GenoMass can automatically generate tables of orderly cut fragments and random cut fragments.

2.3 Results and Discussion

The enzymatic behavior of the benzonase/alkaline phosphatase pair was evaluated by examining progressively larger model oligonucleotides, both single stranded and double stranded. Specifically, single stranded (ss) 12-mer, 20-mer, 21-mer, 80-mer oligonucleotides and double stranded (ds) 12-mer and 21-mer oligonucleotides as well as calf thymus DNA. As indicated below, a "reversed pseudo-combinatorial" approach based on searching a database that consists of the calculated molecular masses of all possible combinations of A, T, C, and G was used to identify the digested products analyzed by LC-MS.

The NarI sequence **-G$_1$G$_2$CG$_3$CC-** and its related sequences which can be found in the β-galactosidase gene of bacteriophage M13mp9 (Tebbs and Romano, 1994) have been documented to exhibit a strong hotspot for -2 frameshift mutation. The NarI sequence was included in several of the selected oligonucleotides. The mutation was induced by a variety of chemical carcinogens that bind primarily to the C8 position of guanine (Bintz and Fuchs, 1990; Burnouf et al., 1989), including N-acetyl-aminofluorene (AAF). AAF adducts are formed when N-acetoxy-2-acetyl-aminofluorene (AAAF), a strong rat liver carcinogen, reacted with DNA. Mutagenesis experiments involving single-adducted plasmids have shown that the mutation frequency induced by dG-AAF residues varies considerably according to the position of the adduct. Only binding to the G$_3$ position of the NarI site leads to high mutation frequencies (Koehl et al., 1989a). Hence, for genetic analysis it is important to determine the mode of enzymatic cleavage of the NarI sequence. In order to illustrate the experimental scheme employed in this study, all samples used in the analysis and their sequences are listed below.

- The following single stranded (ss) 12-mer, 20-mer, 21-mer oligonucleotides and double stranded (ds) 12-mer and 21-mer oligonucleotides were constructed based on the NarI sequence.
  - 12-mer 1, which is the NarI sequence itself

    5'-CCGGCGCCACA-3'

  - 12-mer 2, the complementary strand of the NarI sequence

    3'-TGGCCGCGGTGT-5'

  - Duplex 12-mer, i.e., NarI with complementary strand:

    5'-ACCGGCGCCACA-3'
    3'-TGGCCGCGGTGT-5'

  - 20-mer containing the NarI sequence

    5'-CCCAACACCGGCGCCACAGA-3'

  - 21-mer containing the NarI sequence

    5'-GCCCACAACCGGCGCCACAGA-3'

  - Duplex 21-mer containing the NarI sequence
  -

    5'-ATCCCAACACCGGCGCCACAT-3'
    3'-TAGGGTTGTGGCCGCGGTGTA-5

- A single stranded (ss) 80-mer oligonucleotide was constructed based on all possible tetramer masses that arise from the combination of the four nucleobases A, T, G , C, such that each tetramer contained at least one G. These tetramers were arranged consecutively from the 3' end to the 5' end of the oligonucleotide in an increasing order of the mass of the tetramer fragments:

  5'-AGGGTGGGAAGGGGGCATGGAAAGTTGGAGGCAATGTGG
  CATTGAAG CTTTGGGCCATGCTTGCAGCCTGCCGCCC-3'

- Finally, the enzymatic behavior of the benzonase/alkaline phosphatase pair was further investigated with calf thymus DNA whose sequence is unknown.

The behavior of the enzyme pair was examined in terms of its specificity, the extent of orderly versus random cut, and the size of digested fragments generated. Because of the high incidence of 4-mer fragments in the digested DNA, as observed in the study of the 80-mer and calf thymus DNA (discussed in more detail later), the study focused mainly on the mode of production of 4-mer oligomers, although fragments of greater lengths such as 5-mers and higher were also considered. For the purposes of this book, "*orderly cut*" is defined as the enzyme pair proceeding 4-bases at a time from either the 3' end or the 5' end to generate a 4-mer fragment. On the other hand, "*random cut*" is defined as the process whereby the enzyme pair cuts anywhere in the sequence and generates n-mer fragments, some of which may be isomeric to those produced from an orderly cut or consist of entirely different base composition (referred to here as "*random-cut only*" fragments). Examination of the LC-MS data of standard 4-mer mixtures indicated that the isomers had about equal ionization efficiency. The relative ratios of extracted ion chromatograms could thus be used to quantify the relative distributions of the possible 4-mers resulting from the digestion.

Assuming an orderly cut pattern, the benzonase/alkaline phosphatase enzyme pair was supposed to generate the following 4-mers from the ss 12-mer 1, 5'-ACCGGCGCCACA-3' oligonucleotide: CACA, GCGC, and ACCG. As shown in the chromatogram of **Figure 2.3a**, these 4-mers were formed in high abundance relative to the random cut fragments. For example, note the predominance of GCGC over its "random-cut" isomers of CGGC and CCGG (m/z 1173.8) in the middle panel of **Figure 2.3a** as well as the non-isomeric 4-mer fragments (m/z 1133.8; m/z 1213.8 and m/z 1147.8) shown in **Figure 2.3b**. Thus, based on their respective peak areas, the relative abundances of orderly cut 4-mers appears to be at least one order of magnitude higher than those of their random cut counterparts.

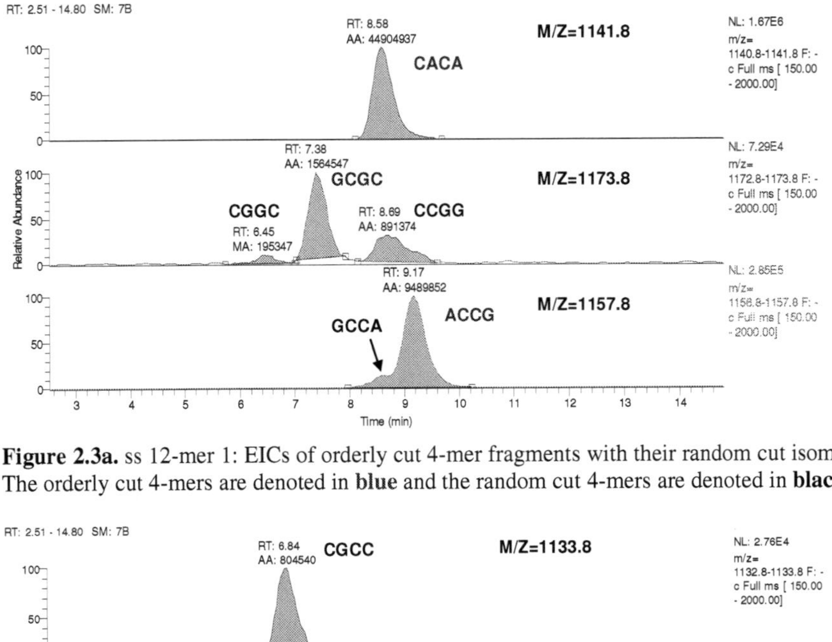

Figure 2.3a. ss 12-mer 1: EICs of orderly cut 4-mer fragments with their random cut isomers. The orderly cut 4-mers are denoted in **blue** and the random cut 4-mers are denoted in **black**.

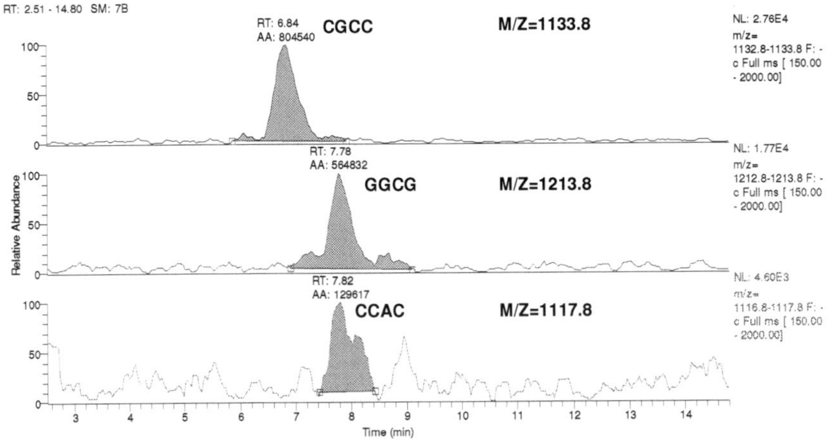

Figure 2.3b. ss 12-mer-1: EICs of random cut only 4-mer fragments.

Because the peaks were chromatographically resolved, the identities of the isomers in the EICs could be readily established by MS/MS as illustrated in **Figures 2.3c and 2.3d.** The results from the MS/MS measurements were in accordance with the peak identifications in **Figures 2.3a** and **2.3b**, thereby confirming the distribution of orderly and random cuts.

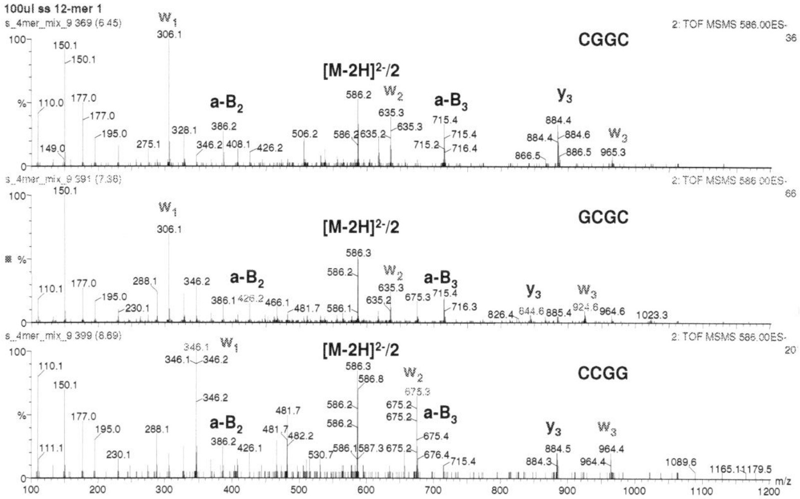

Figure 2.3c. MS/MS of M/Z=1173.8.

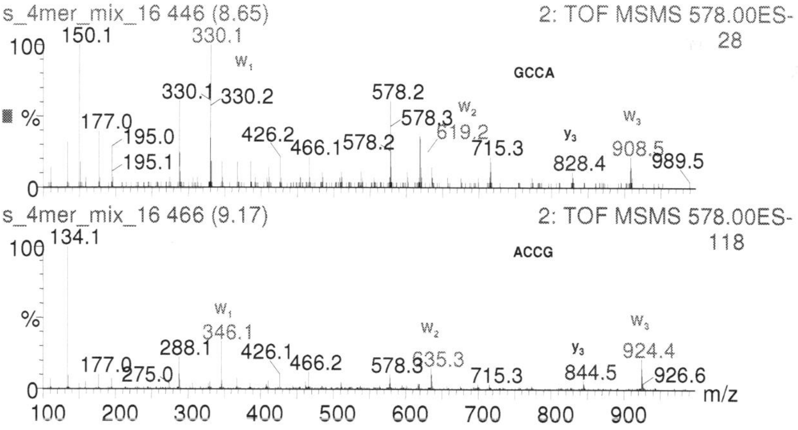

Figure 2.3d. MS/MS of M/Z=1157.8.

The same study was conducted on 12-mer 2 with the sequence: 3'-TGGCCGCGGTGT-5' which is the complementary strand of 12-mer 1. The results were similar to those obtained with the ss 12-mer 1 in that the abundance of the orderly-cut 4-mer fragments was nearly two orders of magnitude higher than that produced from random cuts (data not shown).

Since DNA is double stranded, it was logical to explore whether the 4-mer formation followed the same pattern in the ds 12-mer as in the individual ss 12-mers:

5'-ACCGGCGCCACA-3'.
3'-TGGCCGCGGTGT-5'

The results are presented in **Figures 2.4a** and **2.4b**. **Figure 2.4a** displays the mass chromatogram of the expected orderly-cut 4-mers that may be generated from the upper and lower strands. The results are substantially consistent with those obtained with the individual single strands. From a quantitative perspective, one exception is the higher relative abundance of the GCGC 4-mer (m/z 1173.8) which, in addition to its origination from a normal-cut of the upper strand, may also be derived from random-cut of the lower strand and is thus depicted twice in the figure. As with the ss 12-mers, the relative abundance of the random-cut 4-mers is significantly lower than that of the orderly-cut counterparts, based on the peak areas shown in the EICs (**Figure 2.4b**).

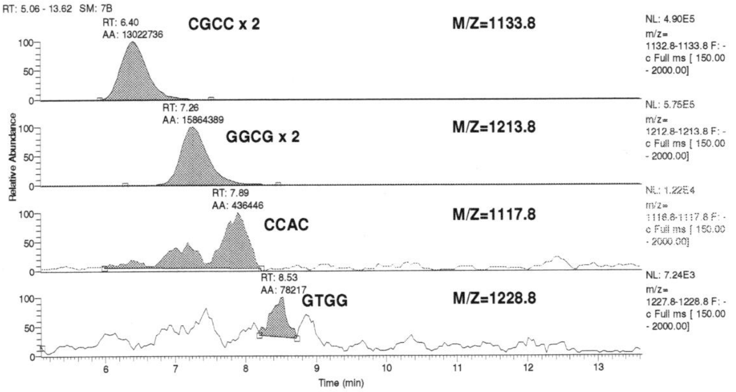

Figure 2.4a. Duplex 12-mer: EICs of orderly cut 4-mer fragments and their random cut isomers, upper strand + lower strand.

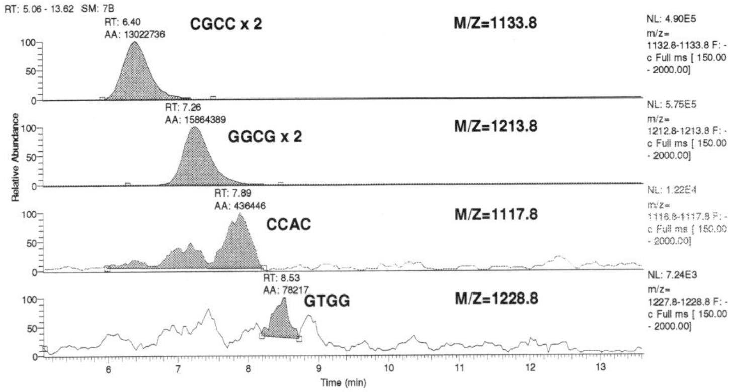

Figure 2.4b. Duplex 12-mer: EICs of random cut only 4-mer fragments, upper strand + lower strand.

Given the apparent higher propensity of the enzyme pair to cleave oligonucleotides into 4-mers in an *"orderly"* fashion, we proceeded to examine progressively longer chain length oligomers since they mimic DNA better. As an example, we discuss next the ss 20-mer sequence, 5'-CCCACAACCGGCGCCACAGA-3', which also contains the NarI sequence. Again, the same general trends were observed as with the 12-mers, with the orderly cuts being predominant as illustrated in **Figures 2.5a** and **2.5b**. The only exception was on m/z=1173.8, where the second peak appears larger. This was due to the contributions from two random cut isomers CCGG + GCGC. However, the abundance of the orderly cut CGGC 4-mer is still slightly higher than each of the random cuts. The isomers' identities were also established by MS/MS (data not shown).

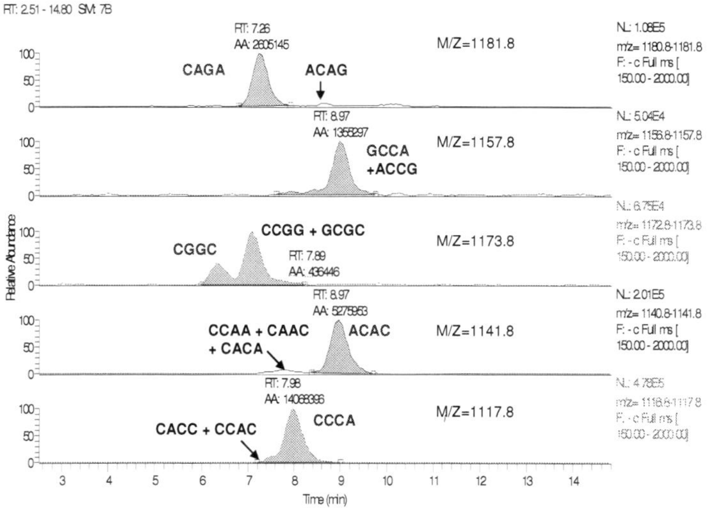

Figure 2.5a. ss 20-mer: EICs of orderly cut 4-mer fragments and their random cut isomers. The orderly cut 4-mers are denoted in **blue** and the random cut 4-mers are denoted in **black**.

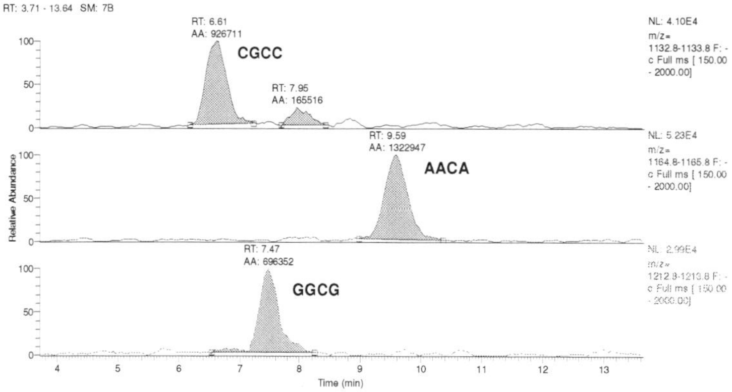

Figure 2.5b. ss 20-mer: EICs of random cut only 4-mer fragments.

Since the sequence of the 20-mer is symmetric, the enzyme pair behavior was further investigated with an asymmetric double stranded 21-mer containing the NarI sequence.

5'-ATCCCAACACCGGCGCCACAT-3'
3'-TAGGGTTGTGGCCGCGGTGTA-5'

All the orderly cut 4-mer fragments from both upper and lower strands were present in the digest as illustrated (marked in blue) in **Figures 2.6a** and **2.6b** which also include the traces from their isomeric random-cut isomers. Consistent with the previous results, the random-cut isomeric 4-mers were present in lower relative abundance. In addition, according to the GenoMass search, two non-isomeric 4-mers derived from the NarI sequence were also detected in relatively significant quantities at m/z 1117.8 and 1228.8 (**Figure 2.6c**). For m/z 1117.8 there were three isomers, CCAC, CACC, CCCA while another three different 4-mers, TGGG, GGTG, GTGG, accounted for the m/z 1228.8 trace. The individual abundances of these random-cut 4-mers were roughly ten-fold lower than those of the orderly cut fragments.

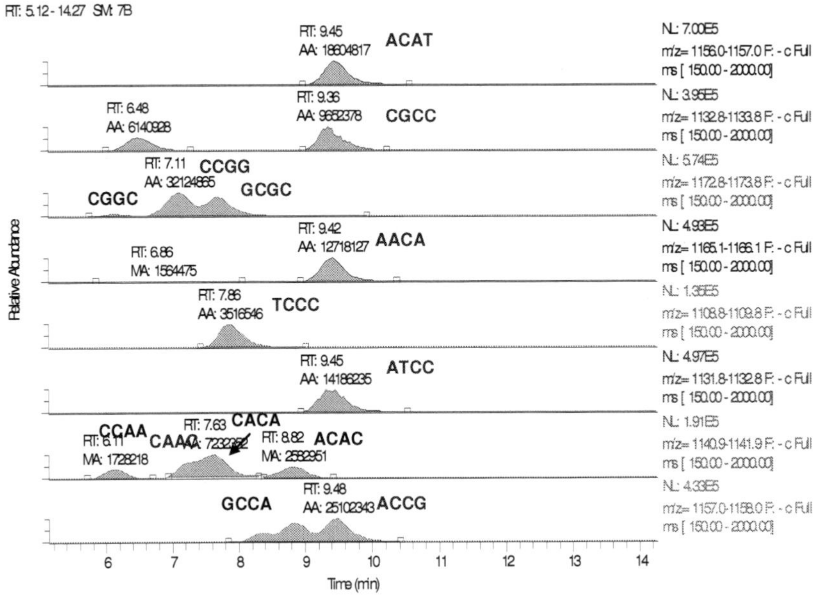

Figure 2.6a. Duplex 21-mer: EICs of orderly cut and random cut 4-mer fragments from upper strand.

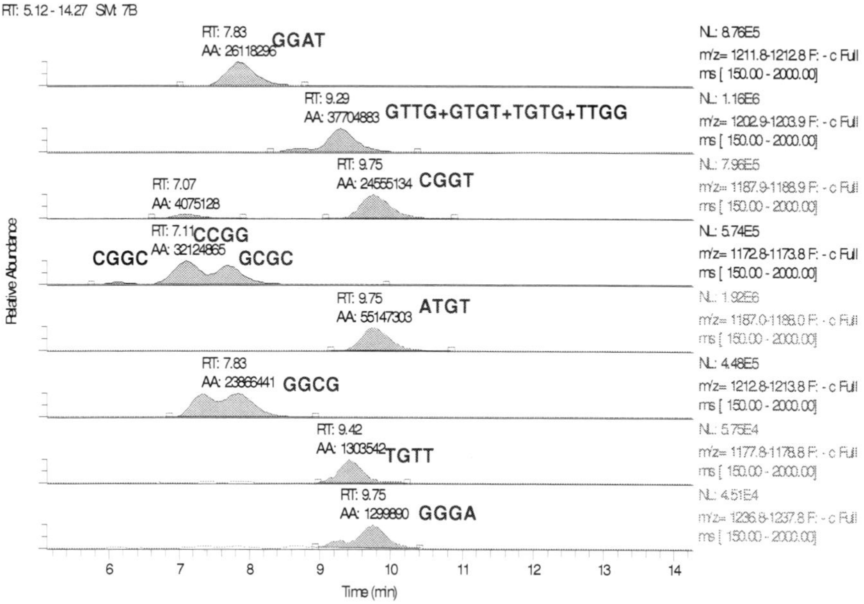

Figure 2.6b. Duplex 21-mer: EICs of orderly cut and random cut 4-mer fragments from lower strand.

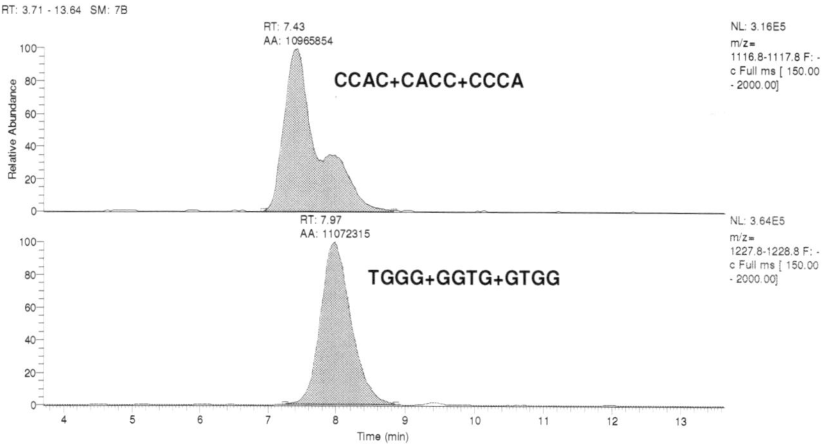

Figure 2.6c. Duplex 21-mer: EICs of random cut only 4-mer fragments.

The same type of analysis was performed next on the ss 80-mer

5'-AGGGTGGGAAGGGGGCATGGAAAGTTGGAGGCAATGTGGCATTGAAG
CTTTGGGCCATGCTTGCAGCCTGCCGCCC-3'.

The results were generally similar to the previous cases, with the orderly cut fragments generated in higher abundance than their isomeric 4-mers formed via random cuts (data not shown). The relationship between the two types of digested fragments is summarized in **Figure 2.7** and, as might be expected, the ratio of orderly to random cut 4-mers decreased with increasing length of the oligonucleotide.

Orderly Cut : Random Cut	12-mer 1	12-mer 2	duplex 12-mer	20-mer	21-mer	duplex 21-mer	80-mer
Ratio	35.52	127.14	6.75	5.17	3.84	3.89	3.89

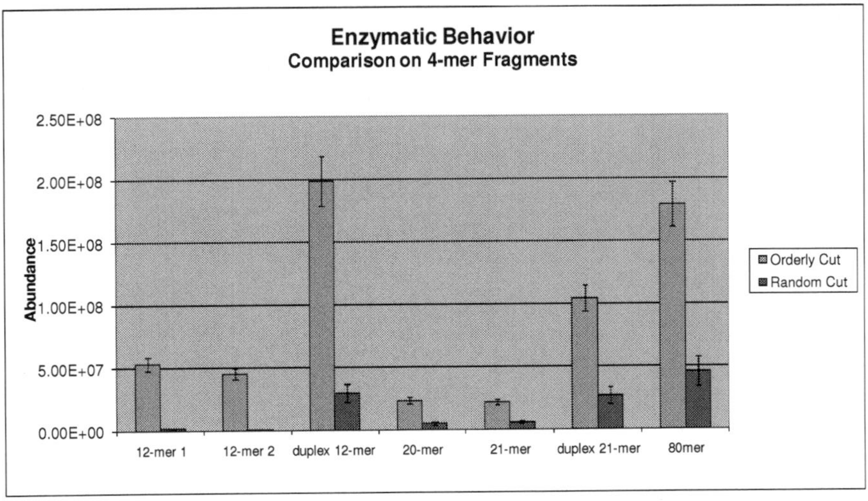

Figure 2.7. Comparison of the EIC peak intensities of orderly cut fragments to random cut fragments.

Because of its longer length, the ss 80-mer represents a more realistic model of what may be encountered in a typical DNA digestion mixture and was thus used to examine in more detail the enzymatic behavior of the benzonase/alkaline phosphatase pair in terms of its overall cleavage pattern and specificity. According to the way the DNA sequence was designed, this 80-mer should yield the following sets of oligomers, assuming enzymatic cleavage from the 3' end of the 80-mer in an orderly manner: 8 dimers with no sequence isomers; 17 trimers; 20 tetramers; and 13 pentamers. On that basis, we should expect to see four isomeric trimers and three isomeric pentamers, but none for the tetramers. The abundance of random cut isomers were not taken into account in the construction of this table and all comparisons are based on the orderly cut fragments and their isomers. All these compounds and their respective ion masses were found by LC-MS analysis and are summarized in **Table 2.1**.

Table 2.1. N-mers present in an oligo 80-mer digested by benzonase/alkaline phosphatase.

N-MERS

Dimers

Base Sequence	**CC**	**AA**	**AG**	**AT**	**GC**	**GG**	**TG**	**TT**
$[M-H]^-$	516	564	580	555	556	596	571	546

Trimers

Base Sequence	**CCC**	**CAT**	**CCG**	**CAA**	**ATT**	**CTG** **TGC** **GCT**	**AGC**	**TTG** **TGT**	**AGT** **ATG**
$[M-H]^-$	805	844	845	853	859	860	869	875	884
Base Sequence	**GGC**	**GAA**	**TGG**	**AGG** **GGA**	**GGG**	**CTT**	**AAT**	**AAA**	
$[M-H]^-$	885	893	900	909	925	835	868	877	

Tetramers

Base Sequence	GCCC	TGCC	AGCC	TTGC	ATGC	GGCC	TTTG	AAGC	ATTG	TGGC
$[M-H]^-$	1355	1370	1379	1385	1394	1395	1400	1403	1409	1410
Base Sequence	AATG	AGGC	TTGG	AAAG	ATGG	GGGC	AAGG	TGGG	AGGG	GGGG
$[M-H]^-$	1418	1419	1425	1427	1434	1435	1443	1450	1459	1475

Pentamers

Base Sequence	**CGCCC**	**CCTGC**	**GCTTT**	**ATGCT**	**CAATG**	**TTGAA** **AAGTT**	**TGCAG** **TGGCA**
$[M-2H]^{2-}$	711	718.5	733.5	738	742.5	750	750.5
Base Sequence	**GGGCC**	**ATGGA**	**GGGGC**	**GGAAG**	**GGGTG**	**GGAGG** **GGGGA**	
$[M-2H]^{2-}$	751	762.5	771	775	778.5	783	

The dimers and trimers were detected as singly charged ions. The tetramers were detected both as singly and doubly charged ions while the 13 pentamers were detected as mainly doubly charged ions. The study was extended to search for 10-mers and above (tables not included here). Although there was evidence that the enzyme produced 11-mer and 12-mer fragments, the abundances of these species were too low to be quantified. The different fragments' elution order corresponded to their size from smaller to larger, which is also in accordance with the increased hydrophobicity of the larger oligonucleotides. The distribution of the different fragments as a function of retention time and mass-to-charge (m/z) ratio is shown in **Figure 2.8**. As expected, the LC separation of the oligonucleotides became increasingly more difficult with increasing size of the fragments, although molecular isomers could be resolved due to the discriminating power of the mass spectrometer.

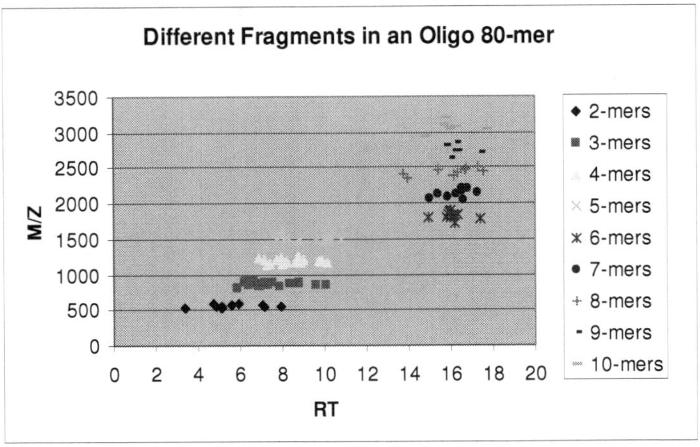

Figure 2.8. 2-D diagram of different fragments in an oligo 80-mer digested by benzonase/alkaline phosphatase.

In order to more accurately compare the relative abundances of the different n-mers, it was necessary to take into consideration any differences in ionization efficiency for the fragments of different sizes. Accordingly, standard mixtures of 3-mers, 4-mers and 5-mers were analyzed using the same LC and MS conditions. It was observed that 4-mers had about 1.5 times higher ionization efficiency than 3-mers and 5-mers, yet the ionization efficiencies for 3-mers and 5-mers were about the same. Therefore, at least for the purposes of these experiments, the differences in ionization efficiency were not sufficiently significant to alter the distribution profile generated from their TICs. Thus, the distribution of the oligomers ranged from 2-mers to 10-mers based on the relative intensity of their extracted ion chromatograms as shown in **Figure 2.9**.

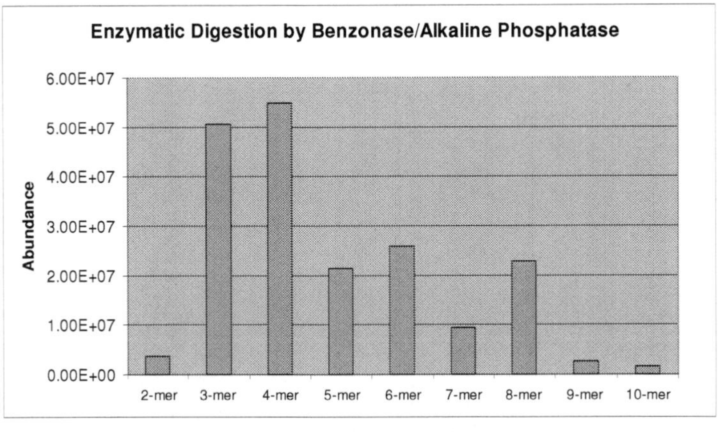

Figure 2.9. Comparison of the abundance of different fragments in an oligonucleotide 80-mer digested by benzonase/alkaline phosphatase.

From these data it may be concluded that the benzonase/alkaline phosphatase pair is capable of cutting oligonucleotides into different fragments from 2-mers to10-mers with a preference for trimers and tetramers, the latter exhibiting the highest yield among all fragments.

Ultimately, the value of the benzonase/alkaline phosphatase enzyme to generate oligonucleotides capable of providing at least some partial sequence related information rests with its ability to handle native DNA. Accordingly, the above series of experiments was repeated with calf thymus DNA whose base sequence was unknown. As expected from the length of the digested fragments, the GenoMass software revealed the presence of numerous possible trimer, tetramer and pentamers. **Figure 2.10** shows the different size fragments present in the LC-MS spectra as clusters of singly charged ions. The doubly charged ions of fragments larger than 6-mers were also present but not labeled here because of their low relative abundance and their m/z overlap with some of the singly charged trimer ions. Multiply charged ions occurring at the same m/z value as singly charged ions can be differentiated by their isotope separation in continuum data mode and/or by their retention times. Their presence was confirmed by ion extraction.

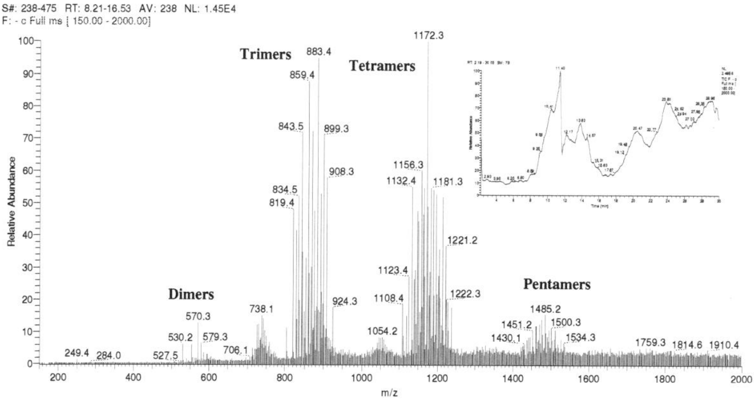

Figure 2.10. Summed mass spectra (time window from 8 min to 16 min) of LC-MS from calf thymus DNA, digested by benzonase/alkaline phosphatase (TIC shown in inset).

In analyzing the calf thymus DNA digest, the GenoMass software automatically generated the six most intense EICs of tetramers (data not shown). As indicated from the EICs, each tetramer profile was likely comprised of several isomers, the sequences of many of which could be established by MS/MS. For the purpose of the present discussion, it should be noted that the GenoMass software demonstrated the ability to easily classify large complex data sets obtained from benzonase/alkaline phosphatase digestion of diverse DNA samples. The enzymatic behavior of this pair was similar for all the standards tested, cleaving DNA non-specifically into different size fragments ranging from 2-mers to 10-mers.

2.4 Conclusions

In order to determine the selectivity of known carcinogens in their reactions with DNA, it is necessary to accurately determine the identity of carcinogen-DNA adducts and pinpoint the exact physical locations of these adducts in the genome. To improve the mass accuracy of measuring DNA molecules, the relatively large DNA fragments (> 30-mer) are usually cut into smaller fragments using different enzymes/nucleases. For the purpose of determining the selectivity of carcinogens, the enzymatic digestion of DNA ought to achieve the preservation of partial DNA sequence information on the digested products, and the continuation of cleavage beyond the location(s) of adduction. Based on these criteria, benzonase and alkaline phosphatase were chosen and their combined enzymatic activity was investigated by LC-MS. An automated search of the mass spectra using a specially designed GenoMass software program, which calculates the molecular masses of all possible combinations of A,G, C, T, has greatly reduced the complexity that might be involved in the identification of digested DNA fragments.

The experiments proved that the benzonase/alkaline phosphatase pair is capable of cutting oligomers, both single and double-stranded DNA, into 2-mers and up to 10-mers. Based on the abundance of these ions, it was observed that tetramers were formed in the highest yield followed by trimers and pentamers. At least for the formation of 4-mers, the cutting took place in a sequential order with the so-called orderly-cut fragments exhibiting much higher relative abundance than their random counterparts. The propensity for an orderly cut decreased with increasing size of the oligomer. It is possible that, when dealing with a small oligomer, the 3' or 5' ends of the oligomer may be more accessible to the enzyme pair to initiate the cleavage in which case production of orderly-cut 4-mers may become more favorable. This selectivity is reduced when the enzymes have to deal with progressively larger and more unwieldy oligomers. In the context of using this enzyme pair to analyze genomic DNA, the ratio of orderly cuts to random cuts can be increased by first digesting the genomic DNA into smaller DNA fragments, which can be easily accomplished by a partial digestion with any non-specific nuclease (e.g. DNase I). Preliminary findings also indicate that this trend also holds in the enzymatic cleavage of AAF-adducted oligomers which should facilitate the determination of modified sites in fragments produced from genomic DNA.

The results of the study also suggest that the benzonase/alkaline phosphatase pair is more non-specific in nature than previously thought. In the past, it was assumed that AT rich regions of DNA are resistant to benzonase cleavage and the enzyme prefers GC-rich sites (Schrader and Linscheid, 1995). Yet the results on the oligonucleotide standards do not support this premise. The enzyme appears to be very much non-specific, which makes it an even better candidate for DNA and DNA-adduct related studies that require the non-specificity of base sequence cleavage.

In conclusion, this study revealed significant information about the enzymatic behavior of the non-specific benzonase/alkaline phosphatase pair using mass spectrometry alone. Further work is currently underway that will facilitate the analysis of more complex digest mixtures. This includes modifications of the GenoMass software to deal with the digests of adducted DNA and interpretation of mass spectral data, and incorporation of a monolithic column with nanoelectrospray to improve chromatographic resolution and sensitivity. It is expected that the methodology employed here could be readily adapted toward the characterization of the enzymatic behavior of different nucleases using LC-MS/MS methods. The further development of the GenoMass software to also include the analysis of carcinogen modified oligonucleotides is discussed in the next chapter of this book.

Supplemental Data:

Table 2.2. Orderly cut fragments and random cut fragments of the studied oligomers – table generated by GenoMass.

NarI 12mer-1	5'-ACCGGCGCCACA-3'							
Orderly Cut	**CACA**	**GCGC**	**ACCG**					
	1142	1174	1158					
Isomers		CCGG CGGC	GCCA					
Random Cut Only	**CCAC**	**CGCC**	**GGCG**					
	1118	1134	1214					
NarI 12mer-2	3'-TGGCCGCGGTGT-5'							
Orderly Cut	**CGGT**	**GCGC**	**TGTG**					
	1189	1174	1204					
Isomers	TGGC	GCCG CCGG						
Random Cut Only	**CGCC**	**GGCG**	**GTGG**					
	1134	1214	1229					
NarI 12mer Duplex	5'-ACCGGCGCCACA-3' 3'-TGGCCGCGGTGT-5'							
Upper strand orderly cut	**CACA**	**GCGC**	**ACCG**					
	1142	1174	1158					
Isomers		CCGG CGGC GCCG	GCCA					
Lower strand orderly cut	**CGGT**	**GCGC**	**TGTG**					
	1189	1174	1204					
Isomers	TGGC	GCCG CCGG CGGC						
Random Cut Only	**CCAC**	**CGCC**	**GGCG**		**GTGG**			
	1118	1134	1214		1229			
NarI 20mer	5'-CCCAACACCGGCGCCACAGA-3'							
Orderly Cut	**CAGA**	**GCCA**	**CGGC**		**ACAC**		**CCCA**	
	1182	1158	1174		1142		1118	
Isomers	ACAG	ACCG	CCGG GCGC		CCAA CAAC CCAC		CACC CCAC	
Random Cut Only	**CGCC**	**AACA**	**GGCG**					
	1134	1166	1214					
NarI 21mer	5'-GCCCAACACCGGCGCCACAGA-3'							
Orderly Cut	**CAGA**	**GCCA**	**CGGC**	**ACAC**		**CCCA**	**GCCC**	**AACA**
	1182	1158	1174	1142		1118	1134	1166
Isomers	ACAG	ACCG	CCGG GCGC	CCAA CAAC CACA		CACC CCAC	CGCC	
Random Cut Only	**GGCG**							
	1214							
NarI 21mer Duplex	5'-ATCCCAACACCGGCGCCACAT-3' 3'-TAGGGTTGTGGCCGCGGTGTA-5'							
Upper strand orderly cut	**ACAT**	**CGCC**	**CCGG**	**AACA**	**TCCC**	**ATCC**	**CAAC**	**ACCG**
	1157	1134	1174	1166	1109	1133	1142	1158

Isomers			GCGC CGGC				CACA CCAA ACAC	GCCA
Lower strand orderly cut	**GGAT**	**GTTG**	**CGGT**	**GCGC**	**ATGT**	**GGCG**	**TGTT**	**GGGA**
	1213	1204	1189	1174	1188	1214	1179	1238
Isomers		GTGT TGTG TTGG	TGGC	CCGG GCCG				
Random Cut Only	**CCAC CACC CCCA**	**TGGG GGTG GTGG**						
	1118	1229						

Chapter 3

GenoMass - A Computer Software for Automated Identification of Oligonucleotide DNA Adducts from LC-MS Analysis of DNA Digests

3.1 Introduction

Advances in the accuracy, sensitivity and efficiency in mass spectrometric technologies and ancillary methods over the past several years have spurred an increased use of the techniques for the characterization of DNA and DNA adducts (Dickman et al., 2002; Vrkic et al., 2000). The general approach toward this goal starts with enzymatic digestion of the DNA followed by LC-MS analysis of the cleaved oligomeric fragments. From the MS data, one may be able to understand the specific enzyme digestion behavior, the oligomeric fragment structure, the binding preference of certain adducts to specific sequences and quantify targeted species at the trace level.

Our current research is focused on the investigation of the selectivity of adduct formation on DNA using ion-pairing reversed phase liquid chromatography mass spectrometry (IP-RP-LC-MS) (Andrews et al., 2004; Dickman et al., 2002; Liao et al., 2007; Oberacher and Huber, 2002; Xiong et al., 2007). As a prerequisite, we first studied the enzymatic behavior of a nonspecific, random endonuclease system, benzonase/alkaline phosphatase (Andrews et al., 2004; Liao et al., 2007). It is envisioned that this enzyme pair may prove to be of value, especially when working with an unknown DNA sequence. The information generated by MS usually needs pre-processing in order to improve the data quality as well as to reduce the data content in order to find the isomeric fragment(s) of interest. A "reversed pseudo-combinatorial" approach is proposed in this paper for this purpose. The method initially considers all possible isomeric combinations of the DNA bases that might yield oligonucleotides of different lengths and searches the LC-MS data for appropriate matches. Thus, by examining each isomer of the combinatorial combinations against the MS data, one can find the isomers and adducts present in the original undigested oligonucleotide or DNA sequence.

The data analysis in studying model oligonucleotides, DNA and their adducts is quite complicated (Barry et al., 1995; Conesa et al., 2005; Fan and Bibikova, 2005; Gaddis et al., 2006; Hail et al., 2004; Lexa and Valle, 2003). For a known model oligonucleotide sequence, it is necessary to find the masses of interest and quantify the peak areas. On the other hand, for an unknown DNA sequence, it is necessary to find and quantify all detectable masses of digested fragments. In MassLynx (Waters/Micromass), LCQ Navigator or XCaliber (ThermoFinnigan) and other data analysis software, this is usually done by searching for the mass and generating an extracted ion chromatogram (EIC). However, with the vast number of fragments produced from a typical DNA digest a program is needed to automate the LC-MS data analysis as efficiently as possible. For an unknown DNA sequence that can be digested into fragments of different sizes, the number of fragments will be $\mathbf{N = 4^n}$, where n stands for the fragment size, or n-mer. Among those fragments there are $\mathbf{M = (4+n-1)!/n!/3!}$ unique masses and for each unique mass there may be numerous positional isomers. Adding another parameter, such as a DNA adduct which contains a carcinogen covalently bound to a site or different sites on the DNA base, introduces a further level of complexity to the data. Without automation, the data analysis and interpretation can be a daunting task. For example, to find a pentamer fragment in the MS data, one needs to compute the pentamer mass, enter the m/z value in a MS tool (such as MassLynx) and look at the output (usually a plot) to inspect whether the pentamer exists. In order to find all the pentamer fragments, this process needs to be repeated $4^5 = 1024$ times, because there are 1024 5-mer combinations of the four bases A, T, G and C. When multiply-charged ions as well as adducts are taken into account, the number of combinations will be several times larger. For larger isomers, the possibilities are enormous and it is impractical if not impossible to process the data manually.

Present software for mass spectrometric characterization of nucleic acids can fulfill a variety of functions but none that addresses the specific tasks described above. For example, **the Mongo Oligo Mass Calculator** developed by McCloskey and coworkers is a handy tool for calculating masses of oligonucleotides, electrospray series, fragments obtained by collision induced dissociation (CID), and fragments of enzymatic digests by endonuclease and exonuclease (Rozenski et al., 1999). The program can also calculate the masses of modified residues. An accompanying tool is **Oligo Composition Calculator** which finds oligonucleotide composition when the user inputs a mass. Mongo Oligo Mass Calculator is by far the most comprehensive program for the analysis of genetic mass spectrometric data with multiple functionalities. **SOS** (Simple Oligonucleotide Sequencer), developed by the same group, is an interactive tool for the de novo determination of short nucleic acid sequences (Rozenski and McCloskey, 1999). The sequence algorithm works by extending from both the 5' (a-B-ions) and 3' (w-ions) ends ion series encoding the complete DNA sequence. Mass ladders are identified by sequentially adding each of the four possible nucleotide masses and searching for the best match of expected ions. The method compares the fragment ion spectra generated by collision-induced dissociation of multiply charged oligonucleotide ions to the predicted m/z values employing established fragmentation pathways from a known reference sequence. Sequences of 5- to 12-mer oligonucleotides were successfully determined. **GeMS** is a user-friendly, advanced commercial software package for gene design developed by Jayaraj and others at KOSAN Biosciences (Jayaraj et al., 2005). The software comprises an integrated suite of programs that automatically perform the following tasks in gene design: restriction site prediction, codon optimization for any expression host, restriction site inclusion and exclusion, separation of long sequences into synthesizable fragments, T_m and stem-loop determinations, optimal oligonucleotide component design and design verification/error checking. The software has been extensively tested in the design and successful synthesis of over 400 kb of genes, many of which exceeds 5 kb in Length. **Links** and **MS2Links** were developed by Young and coworkers at Sandia National Laboratories as a program to analyze mass spectrometric data generated from native, modified, and crosslinked protein and nucleic acid digests (Kellersberger et al., 2004). Links calculates the theoretical crosslinking and modification possibilities for single or multiple protein, RNA and/or DNA sequence(s) given information about the crosslinkers and proteases/nucleases used and the expected amino acid or base modifications. Links then returns putative assignments within a defined mass error threshold for a list of input mass (MH^+) values. MS2Links is an analogous program of Links that is capable of assigning MS/MS peak lists generated from the fragmentation of crosslinked, modified or unmodified peptides, proteins and/or nucleic acids. MS2Links calculates the theoretical MS/MS fragment library given information about the identity of the base ion, crosslinkers (if applicable), desired ion types, and amino acid modifications. It then returns assignments within a defined mass error threshold for the list of input mass (MH^+) values. **GNU polyxmass** is a software framework developed by Filippo Rusconi for mass spectrometric simulations of linear (bio-)polymeric analytes that performs common (bio-)chemical simulations along with simultaneous mass spectrometric calculations for DNA, RNA, saccharides or proteins (Rusconi, 2006). Using the program, the user can 1) define brand new polymer chemistries, 2) perform quick mass calculations using a desktop calculator paradigm, 3) graphically edit polymer sequences and perform (bio-) chemical/mass spectrometric simulations.

While of considerable value in several respects, none of these programs can search through a mass chromatogram, deconvolute the spectra, analyze the peaks, and make the intelligent call on the identity and the sequence of the ion peak based on the parent ion and the fragment ions. This type of data interpretation has to be done manually. The GenoMass software tool presented here is designed to solve these problems. It can automatically calculate ion masses of digestion products of DNA, DNA adducts of various sizes, search for them in complex LC-MS data, list their intensity and plot their EICs. It can also generate prospective n-mer fragments in defined enzymatic cutting patterns. In addition, GenoMass provides specialized data processing and reduction algorithms to increase the the confidence in the results as well as to increase the computation speed. The utility of the software is illustrated here in the investigation of the enzymatic digestion of oligonucleotides and DNA with benzonase/alkaline phosphatase and in the recognition of oligomeric N-acetylaminofluorene (AAF) adducts formed in the reaction of the carcinogen N-acetoxy-2-acetylaminofluorene (AAAF) with DNA (Liao et al., 2007).

GenoMass's main functionalities at the current stage are automation of the data analysis tasks such as ion extraction/plotting and streamlining the complex data. It is particularly handy in finding the adducted oligonucleotide fragments after enzymatic digestion and predicting the mass ions of digestion products of unmodified and modified oligo/DNA fragments. All of the functions are performed on an intuitive, user-friendly program. More functions can be added on top of the ones already in place, according to the needs of the researchers. This "reversed pseudo-combinatorial approach" has not been used in any of the aforementioned software.

3.2 Methods

3.2.1 Experimental Conditions

Sample Preparation

(1) Stock Solution Preparation

The tubes containing **x** nmol (**x** is the actual yield of each standard which varies), ss and ds oligonucleotide standards were centrifuged prior to opening. Then **x** μL DI water was added to the NarI 21-mer, and 80-mer oligonucleotides respectively to create stock solutions of concentration 1 nmol/μL for each sample. The NarI 21-mer was resuspended with **x** μL RNase-Free Duplex Buffer to make a 100 μM solution. These mixtures were all heated to 94 °C for 2 minutes, removed from heat and allowed to cool slowly to room temperature. This resulted in stable, double stranded form of oligomers and they were stored at –20 °C. 1 mg calf thymus DNA was weighed out and 1 mL 10 mM citrate buffer (pH 6) was added to a concentration of 0.1 nmol/μL. All samples were then vortexed down.

(2) AAAF Modification

35.5 μL of 1 nmol/μL 4-mer oligonucleotide standards, NarI 21-mer, and ss 80-mer oligonucleotides were centrifuged down. 355 μg calf thymus DNA was added to 355 μL of 10 mM citrate buffer (pH 6) to a concentration of 0.1 nmol/μL DNA. The above oligonucleotide solutions and DNA were mixed with the AAAF stock solution of 3.55 nmol/μL in acetonitrile in 1:1 molar equivalence. The reaction was allowed to progress at room temperature. The mixtures were then incubated on a heat block at 37 °C for 18 hours.

(3) Enzymatic Digestion

The modified DNA was extracted 3 times with 1 mL ethyl acetate to remove the unreacted AAAF. The aqueous fractions were then dried down using a Speedvac from Savant (Pleasanton, CA) for 2 h at medium heat. The modified oligomers and DNA were then reconstituted in 200 μL of 5 mM Tris/10 mM $MgCl_2$ buffer (pH 8). After being vortexed down, 5 μL benzonase and 5 μL alkaline phosphatase were added to the oligomers and DNA in buffer solution. The mixtures were vortexed well and incubated for 24 h at 37 °C.

(4) Solid Phase Extraction

The digest mixture then went through solid phase extraction (SPE) to concentrate the DNA adducts and remove the unadducted DNA fragments. SPE was performed by using a vacuum manifold fitted with disposable Teflon needles to prevent sample cross-contamination. Unmodified oligomers were washed out while modified oligomers were eluted. The optimized SPE conditions were as follows: (1) Conditioning with 1 mL of methanol. (2) Conditioning with 1 mL of water. (3) Load the samples. (4) Wash with 1 mL 5% (v/v) methanol/water. (5) Elute with 75% (v/v) methanol/water. The adducts-containing fraction was dried down and reconstituted in 100 μL water for analysis.

Chromatographic Conditions

The LC system employed was an Agilent Technologies 1090 HPLC with an autosampler (Wilmington, DE). The HPLC system was controlled by HP ChemStation software running on an IBM-compatible PC. The reverse phase RP-1 PSDVB (polystyrene divinylbenzene) column (2.1 mm x 100 cm x 3.5 μm) was purchased from Hamilton Chromatography (Reno, NV). The column was kept at ambient temperature (25 °C) during analysis. The mobile phase consisted of 25 mM tetraethylammonium bicarbonate (TEAB) as A and 25 mM TEAB with 20% acetonitrile as B. The gradient went from 0%B to 100%B in 20 minutes and was held at 100%B for 5 minutes. The flowrate was 0.2 mL/min and the injection volume was 20 μL.

Mass Spectrometric Conditions

MS data were acquired on a ThermoFinnigan LCQ Classic mass spectrometer (San Jose, CA). The mass spectrometer was tuned on a series of standard oligonucleotides. The samples were analyzed by negative ion electrospray mass spectrometry, first in full scan mode, and subsequently by MS/MS. The MS conditions were as follows: spray voltage: 3.8 kilovolts, capillary voltage: -4 volts, capillary temperature: 200 °C, sheath gas: 25 arbitrary units, tube lens voltage: -25 volts. Data were acquired and processed with Navigator 1.2 software. Mass spectra were recorded in continuum mode.

Oligonucleotides, oligonucleotide-adducts and DNA adducts were first digested to suitable lengths by benzonase/alkaline phosphatase. The digestion products were then subjected to ion-pairing LC-MS and LC-MS/MS analysis. The raw data files were converted to a standard MassLynx raw data file format upon which GenoMass will operate. The data were then processed with several filtering algorithms to reduce the interference and noise as well as to reduce the data. Subsequently, a "reversed pseudo-combinatorial" approach was applied to analyze the data, identify the isomers and extract the information of interest.

In the study of calf thymus DNA, the data obtained with the LCQ MS did not show adequate detection on the modification products, which indicated the LC-MS condition was not sensitive enough to be applied to an in vivo system. Later, we improved the ion pairing reversed phase LC-MS method by switching to nanoLC-μESI with a monolithic column, albeit the solvent composition and ion-pairing reagent were kept the same. The nanoLC system is an Eksigent ExpressLC-100. The LC system was controlled by the MassLynx 4.1 software. The monolithic column (75 μm i.d, 360 μm o.d, 15 cm) was made in house with a PSDVB stationary phase. The flow rate was 200 nL/min. The gradient initially stayed at 98%A, 25 mM TEAB for 3 minutes, then went from 98%A to 98%B, 25 mM TEAB with 20% acetonitrile in 40 minutes. Then the gradient was held at 98%B for 20 minutes. The MS was conducted on a Waters Q-Tof micro system at the following conditions: sample cone voltage at -55 volts, capillary voltage at 1500 volts, extraction cone voltage at 4.5 volts, cone gas at 11 L/hr, desolvation gas at 15 L/hr, source temperature at 100 °C.

3.2.2 Data Pre-processing

The raw data from mass spectrometry may contain much irrelevant information as well as background noise. Proper filtering was employed to make the data more reliable and compact in size. The data preprocessing implemented in the GenoMass software includes background subtraction, calibration correction, noise threshold processing, integration over scan time interval, integration over mass interval, total ion current (TIC) threshold processing, and retention windowing. The following section highlights some of these techniques.

Initially, data from different applications software are converted into standard MassLynx raw data file format by using Data Converter in XCaliber or other data conversion tools such as DataBridge in MassLynx. The original data could come from any common MS system, such as those manufactured by Micromass, Thermo, Applied Biosystems, Agilent etc. Specifically, the following data file formats can be accommodated by the GenoMass program: Xcaliber (*.raw), ICIS (*.dat), GCQ (*.ms), Magnum (*.ms), ANDI (*.cdf), Automas (*.spa), Masslab2 (*.raw), Lasermat (*.*).

Subsequently, a retention window is applied to the raw data so as to reduce the processing time and remove irrelevant information. The start time and stop time of the retention window are normally known and can be configured through the user interface. The next step is to apply a noise filter on the data which are within the retention window. The filter first computes the baseline of the intensity data by a moving-average operator with a 2D triangle weighting-window. It then calculates the signal-to-baseline ratio of each data point and compares them with a preset threshold.

Two criteria must be satisfied in order to declare a raw data point as a valid mass datum. One is that the intensity of the raw data must be over a preset S/N level. Another is that the signal-to-baseline ratio of the raw data must be above a preset ratio threshold. Both noise level and ratio thresholds can be set by the user from the graphic user interface. Raw data not satisfying the above criteria are considered to be noise and are dropped immediately.

The qualified data are used to build a mass list on which each mass (m/z) is unique in value and has a peak intensity associated with it. Since the raw MS acquired data are in terms of m/z values, the data on the mass list are also displayed accordingly. The mass list contains all the expected isomer combinations. GenoMass uses COM objects (.dll) to connect to

Micromass MassLynx software so that all the MassLynx graphic functionalities are available for the user. This means that a user who is familiar with the MassLynx tool does not need to spend any time to learn a new graphic tool. Raw data generated by different machines can be converted to the format that can be processed in GenoMass and MassLynx.

3.2.3 The "Reversed Pseudo-Combinatorial" Approach

In oligonucleotide and DNA analysis by LC-ESI-MS, negative ion detection is normally used due to the phosphate presence in the compounds. For LC-MS analyses employing the most common MS systems of relatively low mass range and/or moderate resolution (e.g. quadrupoles, ion traps), the size of the fragments which can be comfortably characterized is limited to somewhere around 10 to 15-mers (McLuckey et al., 1992; Rusconi, 2006). In the digestion of oligonucleotides or DNA by benzonase/alkaline phosphatase, the fragments observed are typically of short length occurring in the range of 2-mer to 12-mer. According to our LC-MS data, at least for this enzyme system, the occurrence of fragments larger than 10 to 12-mers is rare and their abundances are near or below noise level.

On the basis of the above criteria, this investigation focused on the consideration of MS data representative of 2-mer to 12-mer fragments for the purpose of their detection, identification and quantification. In order to achieve those objectives, a "reversed pseudo-combinatorial" approach is proposed here. This approach works from the expected results back to the MS data (thus, the term "reversed"). The expected results are the isomers belonging to an isomer library which is comprised of all the 2-mer to 12-mer isomers, i.e., all the combinations of the four bases A, T, G and C with segment lengths from 2 to 12 bases (hence our use of the term "pseudo-combinatorial"). For each isomer in the library, the program searches through the mass list obtained in the previous step to see if there is a mass value identical to the computed isomer mass. If a match is found, the presence of the isomer in the original oligonucleotide or DNA sequence is claimed. The isomer information, including its name, structure, mass and peak intensity (ion abundance), are registered on the result list. This procedure is repeated until all the isomers in the isomer library have been processed.

For small oligomers, the isomer library is generated in the software on the fly while for relatively larger isomers (> 8-mer), the library is pre-generated and saved in a database in order to speed up the computation. **Figure 3.1** demonstrates an example of the generation and display of a tetramer library.

4-mer	Sequence		M	M-H	M-AAF-H	M-2H/2	M-AAF-2H/2
1095	CCCC	1	1094.8	1093.8		540.4	
1110	TCCC CTCC CCTC CCCT	4	1109.8	1108.8		553.9	
1119	ACCC CACC CCAC CCCA	4	1110.0	1117.8	[illegible]	550.4	[illegible]
1125	TTCC TCTC TCCT CTTC CTCT CCTT	6	1124.8	1123.8		561.4	
1134	ATCC ACTC ACCT TACC TCAC TCCA CATC CACT CTAC CTCA CCAT CCTA	12	1133.8	1132.8	[illegible]	565.9	676.4
1135	GCCC CGCC CCGC CCCG	4	1134.8	1133.8	1354.8	566.4	676.9
1140	TTTC TTCT TCTT CTTT	4	1139.8	1138.8		568.9	
1143	AACC ACAC ACCA CAAC CACA CCAA	6	1142.8	1141.8	[illegible]	570.4	[illegible]
1149	ATTC ATCT ACTT TATC TACT TTAC TTCA TCAT TCTA CATT CTAT CTTA	12	1148.8	1147.8	[illegible]	573.4	[illegible]
1150	TGCC TCGC TCCG GTCC GCTC GCCT CTGC CTCG CGTC CGCT CCTG CCGT	12	1149.8	1148.8	1369.8	573.9	[illegible]
1155	TTTT	1	1154.8	1153.8		576.4	
1158	AATC AACT ATAC ATCA ACAT ACTA TAAC TACA TCAA CAAT CATA CTAA	12	1157.8	1156.8	1377.8	577.9	[illegible]
1159	AGCC ACGC ACCG GACC GCAC GCCA CAGC CACG CGAC CGCA CCAG CCGA	12	1158.8	1157.8	1378.8	578.4	[illegible]
1164	ATTT TATT TTAT TTTA	4	1163.8	1162.8	[illegible]	580.9	[illegible]
1165	TTGC TTCG TGTC TGCT TCTG TCGT GTTC GTCT GCTT CTTG CTGT CGTT	12	1164.8	1163.8	[illegible]	581.4	[illegible]
1167	AAAC AACA ACAA CAAA	4	1166.8	1165.8	[illegible]	582.4	[illegible]
1173	AATT ATAT ATTA TAAT TATA TTAA	6	1172.8	1171.8	[illegible]	585.4	[illegible]
1174	ATGC ATCG AGTC AGCT ACTG ACGT TAGC TACG TGAC TGCA TCAG TCGA GATC GACT GTAC GTCA GCAT GCTA CATG CAGT CTAG CTGA CGAT CGTA	24	1173.8	1172.8	1393.8	585.9	[illegible]
1175	GGCC GCGC GCCG CGGC CGCG CCGG	6	1174.8	1173.8	1394.8	586.4	[illegible]
1180	TTTG TTGT TGTT GTTT	4	1179.8	1178.8	1399.8	588.9	[illegible]
1182	AAAT AATA ATAA TAAA	4	1181.9	1180.9	1401.9	590.0	[illegible]
1183	AAGC AACG AGAC AGCA ACAG ACGA GAAC GACA GCAA CAAG CAGA CGAA	12	1182.8	1181.8	1402.8	590.4	[illegible]
1189	ATTG ATGT AGTT TATG TAGT TTAG TTGA TGAT TGTA GATT GTAT GTTA	12	1188.8	1187.8	[illegible]	593.4	703.9
1190	TGGC TGCG TCGG GTGC GTCG GGTC GGCT GCTG GCGT CTGG CGTG CGGT	12	1189.8	1188.8	[illegible]	593.9	704.4
1191	AAAA	1	1190.9	1189.9	1410.9	594.5	[illegible]
1198	AATG AAGT ATAG ATGA AGAT AGTA TAAG TAGA TGAA GAAT GATA GTAA	12	1197.9	1196.9	1417.9	598.0	708.5
1199	AGGC AGCG ACGG GAGC GACG GGAC GGCA GCAG GCGA CAGG CGAG CGGA	12	1198.8	1197.8	1418.8	598.4	708.9
1205	TTGG TGTG TGGT GTTG GTGT GGTT	6	1204.8	1203.8	1424.8	601.4	711.9
1207	AAAG AAGA AGAA GAAA	4	1206.9	1205.9	1426.9	602.5	713.0
1214	ATGG AGTG AGGT TAGG TGAG TGGA GATG GAGT GTAG GTGA GGAT GGTA	12	1213.9	1212.9	1433.9	606.0	716.5
1215	GGGC GGCG GCGG CGGG	4	1214.8	1213.8	1434.8	606.4	716.9
1223	AAGG AGAG AGGA GAAG GAGA GGAA	6	1222.9	1221.9	1442.9	610.5	721.0
1230	TGGG GTGG GGTG GGGT	4	[illegible]	1228.9	1449.9	614.0	724.5
1239	AGGG GAGG GGAG GGGA	4	1238.9	1237.9	1458.9	618.5	729.0
1255	GGGG	1	1254.9	1253.9	1474.9	626.5	737.0

Figure 3.1. The tetramer library of an unknown oligonucleotide sequence with AAF-adducts. The formula for calculating MW of anhydrous oligonucleotide sequence is: Mass = S (individual base mass) – 63.980 + 2.016. The Mass of AAF = 221.09.

The above considerations are applicable for an unknown oligonucleotide or DNA sequence. For a known oligonucleotide sequence, however, the library is generated differently to improve the computing efficiency. The sequence is first cut into n-mers orderly or randomly from 5'- end to 3'- end and then from 3'- end to 5'- end. The cut segments are sorted according to their masses.

In certain cases, the user may be interested in a specific group of isomers. For example, AAAF adduction occurs primarily on the G-base. In this case, the search may be limited to identifying fragments which have at least one G-base. To serve this purpose, a sub-library is generated on the fly by simply re-scaling the complete library according to the user's specification. To find fragments which contain a covalently bound adduct, the isomer library is modified by adding the mass of the adduct to each isomer mass of the library. Similarly, to find doubly-charged fragments or even adducted doubly-charged fragments, the isomer library is modified accordingly. The modified library is then used to search for the isomer presence in the MS data. Due to the convenient structure of the "reversed pseudo-combinatorial" approach, it is relatively easy to adapt the software to those application variations.

The input and basic interface settings are self-explanatory. As shown in **Figure 3.2**, first, the user loads the raw data file from the main window by clicking the button right next to the Raw Data File box. If the background needs to be subtracted, load the background data file as well. Click the Subtract button to conduct the background subtraction process. Then, if it is known, enter the gene sequence from which the raw data was generated. For an unknown gene sequence, just leave the Gene Sequence box blank. Then, enter the criteria to describe what fragments should be searched in the data. For example, one may want to see if there are any 4-mer fragments in the data which contains G- with AAF adduct. One just enters 4 in Isomer Type box, G in Isomer Containing box and 221.09 in the Adduct box.

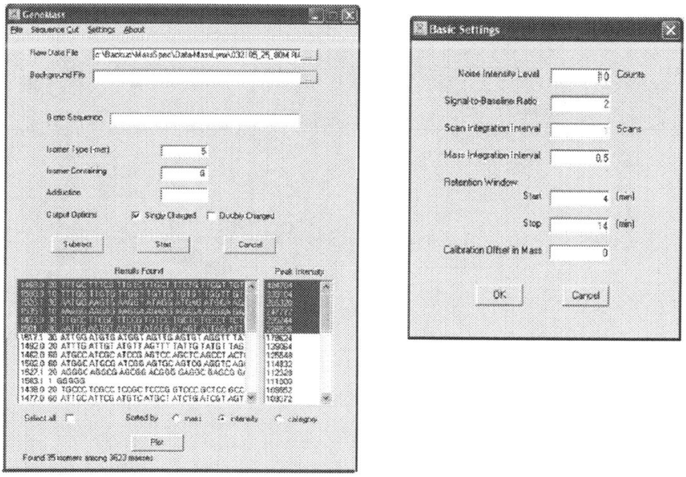

Figure 3.2. The interfaces used by GenoMass to search for the 6 most abundant 5-mers in the digest of oligo 80-mer.

Some parameters related to data processing should be set as well in the Basic Settings window. When those are done, a click of the "Start" button initiates the data analysis process. The output box lists all the found fragments and the corresponding peak intensity values. The results may be sorted according to mass, intensity or category by checking the options below the output box. They can be exported to an Excel program for further analysis by using the Save utility under the File menu. Since the GenoMass software is connected to MassLynx by COM objects, one can easily view the graphic presentations of the EIC and spectra of those found isomers by simply clicking the Plot button of the interface.

3.2.4 Software Implementation

The software utilizes client-server architecture for implementing the "reversed pseudo-combinatorial" approach. As shown in **Figure 3.3**, the system consists of a client, a search engine server and an isomer library server. The client first conducts the data pre-processing and data reduction on the raw MS data. This generates a mass list consisting of a collection of unique mass-intensity pairs. The search engine uses this list as its database. Upon receiving a query from the client, it searches the database to see if there is a match within a certain mass range and returns the found mass-intensity pair accordingly. The default range is ±0.1 (m/z) and can be changed from the "Basic Settings" window under the item "Integration Mass Interval" (**Figure 3.5**).

The isomer library server generates an isomer mass table as described previously and returns it to the client whenever requested. The client examines each isomer in the mass table and queries the search engine to look for a match in the MS data. When a match is found, the corresponding isomer and its related information are registered on the output list.

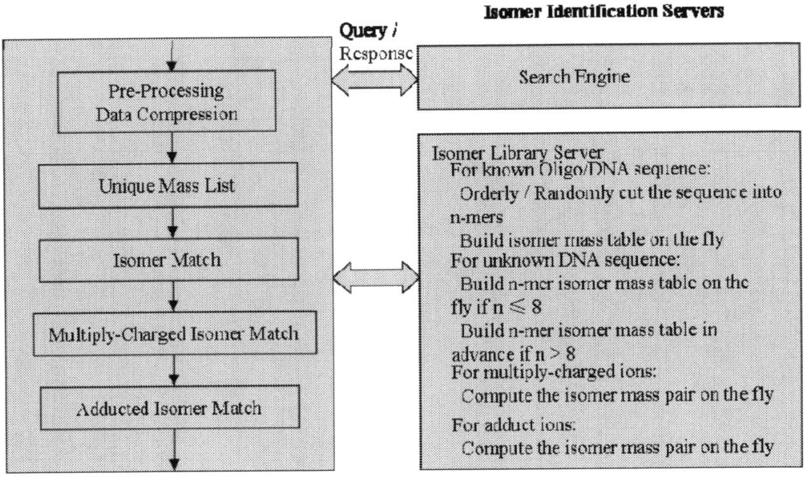

Figure 3.3. Client-server architecture for implementing the "reversed pseudo-combinatorial" approach.

GenoMass software was written in Visual Basic 6.0 under Windows XP. The installation packaging was done with Inno Setup 5.2.3 software. The recommended minimum requirements to run this software are 1.0GHz CPU, 512KB memory and 1GB hard drive, running Windows XP or Windows 2000.

When the server first starts, it creates the isomer-mass tables for all n-mers, where n is from 2 to 12. It then waits for the client to query. When it gets a query, it searches the corresponding isomer-mass table and replies to the client with the findings.

To save memory space, the isomer-mass tables for adducts and multi-charged ions are generated on-the-fly. They are all based on isomer-mass tables which have already been created when the server starts. Therefore, those on-the-fly calculations do not take much computing power. Eventually, the GenoMass functions can also be implemented on a Mac platform. The following charts detail the work flow of GenoMass.

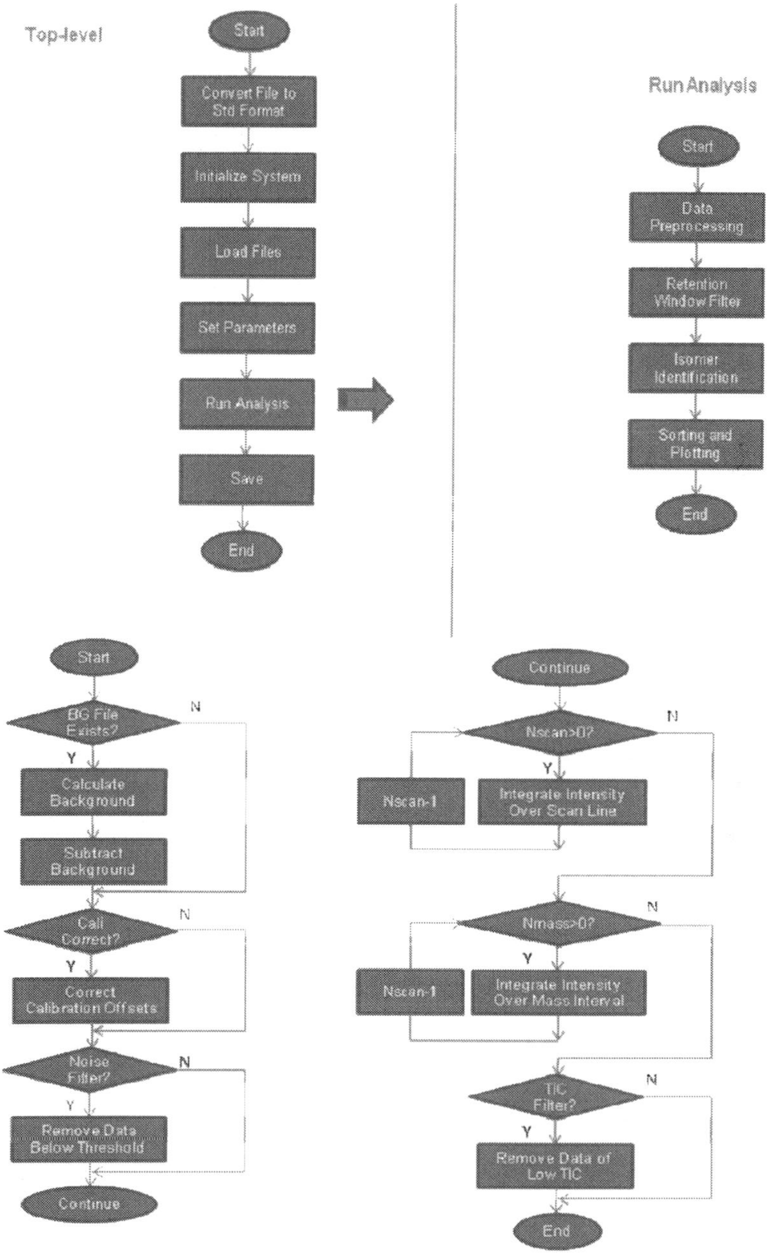

Top-level
Start
Convert File to Std Format
Initialize System
Load Files
Set Parameters
Run Analysis
Save
End
Run Analysis
Start
Data Preprocessing
Retention Window Filter
Isomer Identification
Sorting and Plotting
End
Start
BG File Exists?
N
Y
Calculate Background
Subtract Background
Cali Correct?
N
Y
Correct Calibration Offsets
Noise Filter?
N
Y
Remove Data Below Threshold
Continue
Continue
Nscan>0?
N
Y
Nscan-1
Integrate Intensity Over Scan Line
Nmass>0?
N
Y
Nscan-1
Integrate Intensity Over Mass Interval
TIC Filter?
N
Y
Remove Data of Low TIC
End

Retention Window Filter

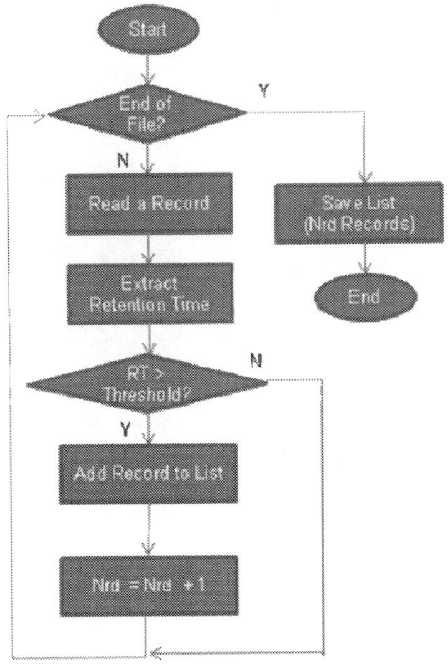

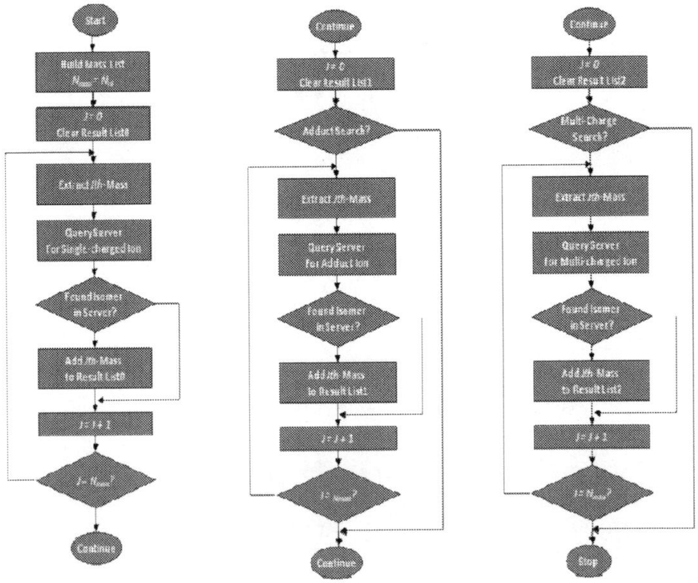

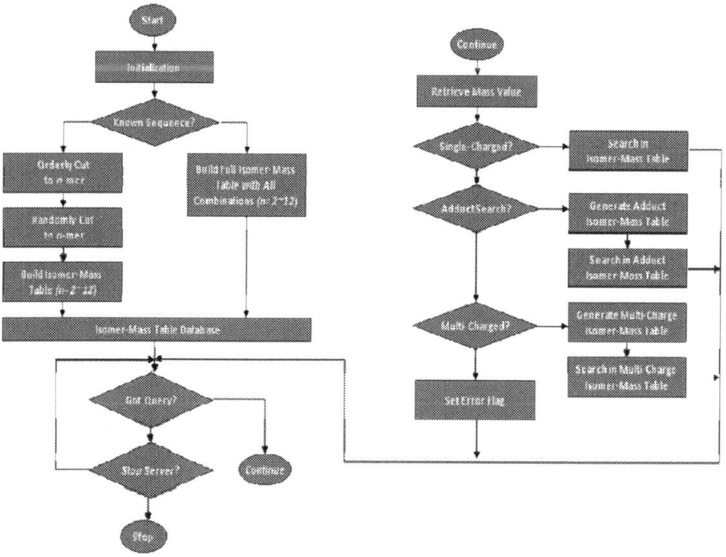

Figure 3.4. Flow charts of GenoMass software.

3.3 Results and Discussion

As an example of the utility of the GenoMass software, we present in this section results from the processing of LC-MS data of enzymatically digested model DNA systems. An enzyme cocktail consisting of benzonase/alkaline phosphatase, which is known to cleave DNA in a non-specific manner, was used in all these studies. Since the ultimate objective of our work is the application of the GenoMass software to the study of carcinogen-modified oligonucleotides, the 12-mer NarI gene sequence which is a known mutational target for the carcinogen N-acetyl-N-acetoxyaminofluorene (AAAF) was incorporated in the model systems. Initially, the software was tested with the enzymatic digests of two unmodified oligonucleotides, a double stranded 21-mer duplex and a single stranded 80-mer. Following confirmation of the feasibility of the general approach, both the 80-mer and calf thymus DNA (ctDNA), whose structure is unknown, were reacted with AAAF and the software was used to search the LC-MS profiles of the respective digests for oligonucleotide adducts generated by enzymatic cleavage. The model sequences tested are shown in **Table 3.1** with the NarI gene sequence segment highlighted in blue and the hotspot in red.

Table 3.1. Sequences studied in the investigation of enzyme behavior and the selectivity of DNA-adduct formation.

Oligonucleotide ID	**Sequence**	**Mass** (da)
NarI 12-mer-1	5'-ACCGGCGCCACA-3'	3598.6
NarI 21-mer Duplex	5'-ATCCCAACACCGGCGCCACAT-3' 3'-TAGGGTTGTGGCCGCGGTGTA-5	12898.2
Oligo 80-mer	5'-GGGGAGGGTGGGAAGGGGGCATGGAAAGTTGGAGGCAATG TGGCATTGAAGCTTTGGGCCATGCTTGCAGCCTGCCGCCC-3'	25047.1
Calf thymus DNA	Unknown	

1. Benzonase/Alkaline Phosphatase Digests of Unmodified Oligonucleotides.

In earlier studies of oligonucleotide digests by benzonase/alkaline phosphatase it was reported that, in the generation of 4-mer fragments from short DNA segments, the enzyme pair proceeded by cleaving four bases at a time from either the 3'- end or the 5'- end (Liao et al., 2007). For the purposes of this study, we define this pattern as "*orderly cut*". On the other hand, "*random cut*" is defined as the process whereby the enzyme pair cuts anywhere in the sequence and generates n-mer fragments, some of which may be isomeric to those produced from an orderly cut or consist of entirely different base composition (referred to here as "*random-cut only*" fragments). In the case of the benzonase/alkaline phosphatase digestion of the NarI 21-mer Duplex, which is an asymmetric sequence

5'-ATCCCAACACCGGCGCCACAT-3'
3'-TAGGGTTGTGGCCGCGGTGTA-5'

orderly cut and random cut fragments were found in the experimental data, from both the upper strand and the bottom strand and their abundances were compared. GenoMass computed the library of fragments from all possible cuts, as shown in **Figure 3.4**. This information was then used to search the LC-MS data for the presence of such fragments in the digested sample.

In this case, if cut orderly from either 3'- end or 5'- end, ten 4-mer fragments with different base sequences were generated. However, two sets of fragments were isomers: CAAC and CACA are both of m/z 1141.8; CCGG and GCGC are both of m/z 1173.8. If cut randomly only, meaning starting anywhere except from the beginnings of the sequences, then eight 4-mer fragments were generated. There were also two sets of isomers: CCAC, CACC, CCCA are all of m/z 1117.8; ACAC and CCAA are both of m/z 1141.8. Note the display in **Figure 3.5** of the m/z or the number of occurrences for the fragment(s) of a given m/z value (isomer counts) derived from either orderly cut or random cut. For example, a frequency number of "2' is given for the isomer counts of m/z 1141.8, when considered in terms of an orderly cut. This means that there were two fragments of such m/z value among all the orderly cut fragments. Similarly, we found 7 fragments of this m/z among all the randomly cut results. The exclamation mark "!" indicates that this m/z appeared in both orderly and randomly cut results.

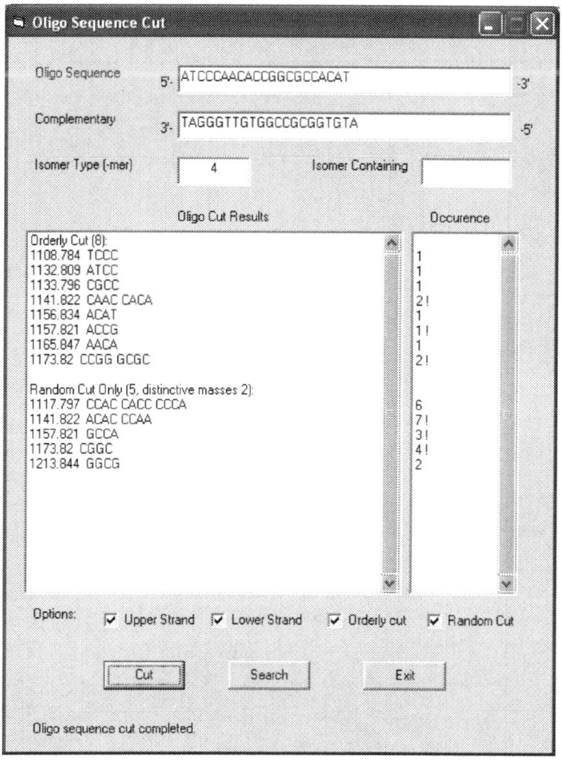

Figure 3.5. The 4-mer enzymatic digestion products of the NarI 21-mer duplex, categorized as orderly and random cut fragments.

The oligo 80-mer presented an opportunity to test the effectiveness of the software and the action of the enzyme in a more realistic and challenging setting than that of the shorter 21-mer. GenoMass searched the LC-MS profile for all possible 2-mers to 12-mers that could be produced from this oligonucleotide of known sequence and listed their intensities. The program then found and compared all the possible ion masses of significant abundance up to 10-mers and determined that the enzyme system cut the 80-mer predominantly into 3-mers and 4-mers and progressively to a lesser extent to the larger 5-mers, 7-mers, etc. In addition, the GenoMass search revealed that there was a higher abundance of orderly-cut than random-cut 4-mer fragments (Liao et al., 2007). As an example, **Figure 3.6** summarizes the output generated by the GenoMass software in the search for 5-mer fragments and the reconstructed ion chromatogram plots of the six most abundant 5-mers in the digested sample.

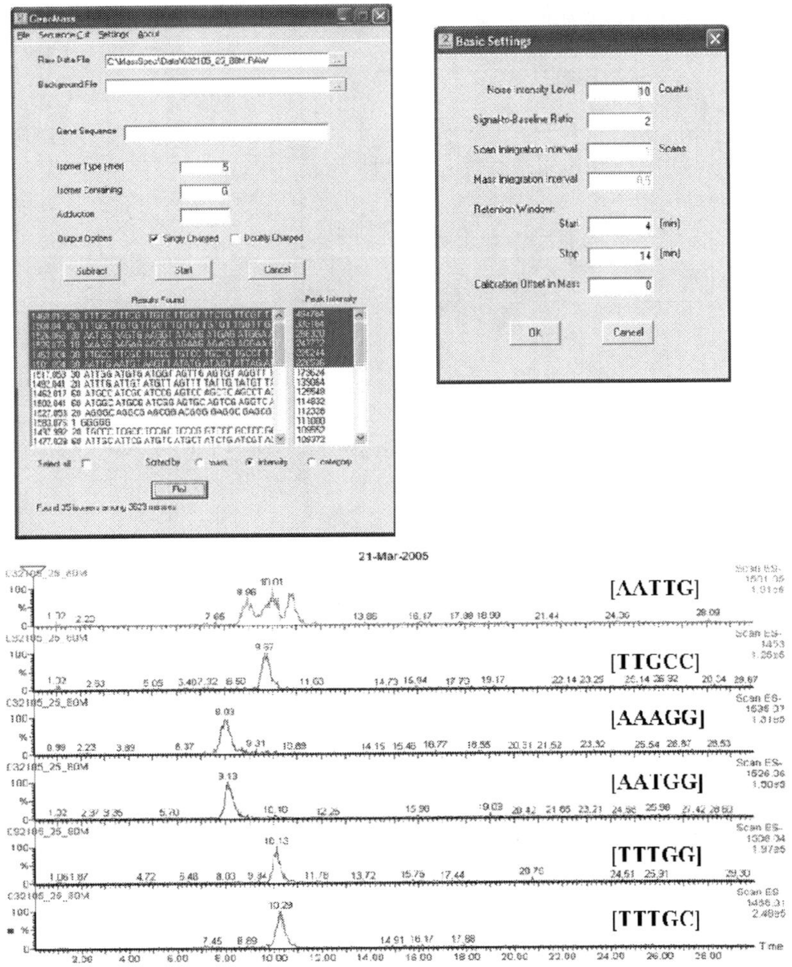

Figure 3.6. EICs of identified the 6 most abundant 5-mers in the digest of oligo 80-mer. Automatically generated and plotted by GenoMass. Brackets indicate base composition rather than sequence. [**x**] denotes the compositional isomers.

<u>2. Recognition of AAF-oligonucleotides in Modified DNA</u>

Of paramount importance in our work is the development of the capacity to recognize oligonucleotide as opposed to monomeric DNA adducts in order to develop an understanding of the sequence dependence in the reaction of carcinogens with DNA bases. Accordingly, the GenoMass software was used to analyze the LC-MS data of the benzonase/alkaline phophatase digest of both the AAF-modified 80-mer and calf thymus DNA. The enzymatic digestion generated complex LC-MS data marked by the presence of a broad range of concentrations of AAF-modified fragments produced from all possible combinations of the four nucleobases A, T, G, C. In this specific case, since the favored target of AAAF is the guanine base, the software was directed toward the search of oligonucleotide adduct masses containing at least one G-unit. For example, a search for G-containing AAF 5-mer fragments from the 80-mer and the calf thymus DNA was conducted and their EICs were selected and plotted with a click of the mouse as shown in **Figures 3.7** and **3.8**, respectively.

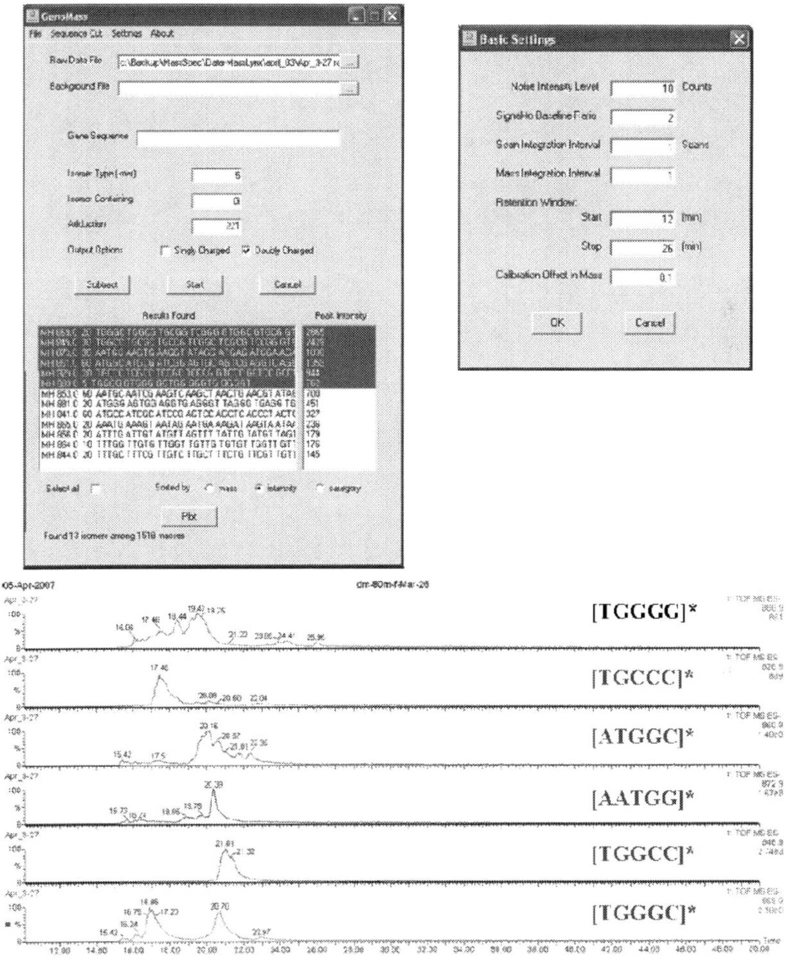

Figure 3.7. EICs of the six most abundant AAF-modified 5-mers identified in the digest of oligo 80-mer following exposure to AAAF. Automatically generated and plotted by GenoMass. [**x**] denotes the compositional isomers.

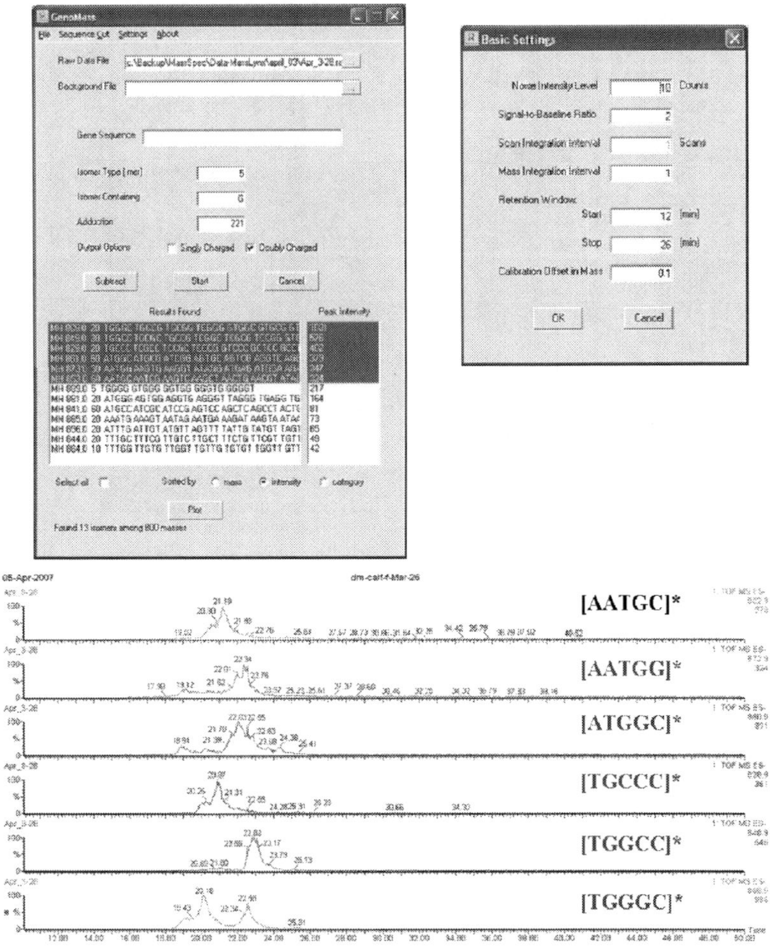

Figure 3.8. EICs of identified six most abundant AAF-modified 5-mers in the digest of calf thymus DNA following exposure to AAAF. Automatically generated and plotted by GenoMass. A 2-minute offset in retention times of adducts common to **Figures 3.6** and **3.7** is due to solvent delay. [**x**] denotes the compositional isomers.

In comparing the results from the digestion of the adducted 80-mer and ctDNA, it was interesting to find that five out of the six most abundant AAF-5-mer masses (m/z) were common to both systems. The data are summarized in **Table 3.2** and partially denoted in the chromatograms of **Figures 3.7** and **3.8**. While presently the search generated information strictly on the base composition of the adducts, it should be noted that full sequence can be obtained by MS/MS in a subsequent LC-MS analysis.

Table 3.2. Comparison of the 6 most abundant modification products (doubly charged ions) of the 5-mer fragments [**x**] found in the ss oligonucleotide 80-mer and the ds calf thymus DNA. The blue colors represent the same composition found in both samples. [**x**] denotes the compositional isomers (as shown in **Figure 3.6** and **3.7**).

$[M-2H]^{2-}$ (da)	**828.9**	**860.9**	**872.9**	**848.9**	**868.9**	**888.9** **852.9**
80-mer	TGCCC	ATGGC	AATGG	TGGCC	TGGGC	TGGGG
DNA	TGCCC	ATGGC	AATGG	TGGCC	TGGGC	AATGC

Ultimately, the ability to relate sequence selectivity of adduct formation in specific domains of DNA or of short oligomers produced in the partial digestion of the DNA would depend on the ability of the software to reconstruct chromatograms of longer fragments. However, when the length of the fragments increases, so does the complexity of the isomer library. Moreover, due to the nature of ESI-MS, larger fragments tend to produce a significantly higher abundance of doubly charged as opposed to singly charged ions. As a consequence, the database is much larger and requires far more computing power. The utility of the GenoMass software to screen the data for longer sequences is demonstrated in the search for 9-mers as doubly charged ions and is illustrated in the chromatogram of **Figure 3.9**, which shows the abundant presence of the 9-mers in the digest of calf thymus DNA reacted with AAAF.

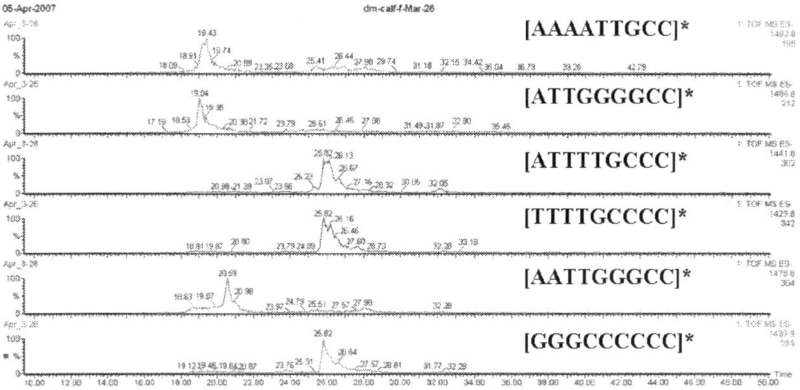

Figure 3.9. EICs of identified six most abundant AAF-modified 9-mers in the digest of calf thymus DNA following exposure to AAAF. Chromatograms were automatically generated and plotted by GenoMass. [**x**] denotes the compositional isomers.

3.4 Conclusions

The "reversed pseudo-combinatorial" approach based GenoMass software is a powerful computer program that can simplify and automate the data analysis process in genomic research. It can be utilized towards the investigation of enzyme behavior and is also directly applicable towards the streamlining of the complex LC-MS data generated in studies associated with the characterization of DNA adducts.

Upon consideration of the length of fragment(s) cut by a given enzyme, one enters the relevant isomer and mass composition into a dialog box, defines the threshold intensity for the selected ion masses and sorts the LC-MS data in a variety of ways. For a known oligomer or DNA sequence with a defined enzymatic cutting pattern, GenoMass can automatically generate corresponding masses of all digestion products. It then searches through the LC-MS raw data file for the masses, outputs their m/z values and their intensities and plots EICs of these found masses. For an unknown oligomer or DNA sequence, GenoMass can automatically search for all masses of possible enzymatic digestion products, output their m/z and intensity and can plot their EICs. When searching for oligonucleotide adducts, the software incorporates the carcinogen mass into the search and is then able to map the base sequences and compositions of the modified oligos. In studies of genomic DNA, the use of this "reversed pseudo-combinatorial approach" can possibly postulate sequence relevant information and indicate sequential preference of adduction. The computer program automates the processing of data that are too complex to be accomplished by manual analysis currently performed with most mass spectrometry application software such as MassLynx and LCQ Navigator. The program is especially useful when handling complex DNA data whose sequence is unknown.

In a broad sense, GenoMass is potentially high throughput computer software capable of fast and efficient data mining of complex mass spectrometric data for nucleic acid analysis. It draws some analogy from software used in proteomic research with the "bottom up" approach yet is quite different due to the difference in the makeup of proteins and nucleic acids. While there are 20 amino acids in peptide and protein samples, there are only four (five) nucleotides in DNA (RNA) samples. While a priori it would appear that the development of software for the analysis of spectral data from genetic samples would be relatively simpler than for proteins, there are actually fewer than a dozen popularly used software in this arena. This is in contrast to the proteomics area where there are nearly 600 different data analysis software available. The reason behind it is that mass spectrometry itself has been used far less in nucleic acid analysis than in proteomics, as array and chip based technologies are more widely employed. However, mass spectrometry has been gaining a stronger footing in genomic research in recent years with the advancement in instrumentation. Particularly in the area of nucleic acid modification, mass spectrometry now can make a much bigger contribution with improved sensitivity, mass accuracy and dynamic range.

Although GenoMass is at its very initial stage, the functionalities are very different from the few existing software for mass spectrometric characterization of nucleic acids. In combining some of the features in programs such as Mongo and SOS, i.e. the MS/MS sequencing and sequence assignment based on both molecular ion and fragment ions, GenoMass can over time evolve into a full functional data analysis tool for mass spectrometric characterization of genomic samples including oligonucleotides, DNA, RNA and their modification products and be made available to the public potentially as an open source software.

In the near future, acquisition of data with higher mass resolution will aid in narrowing the search window and introduce additional selectivity to the GenoMass reconstruction process by improving the quality and accuracy of the chromatographic data. Moreover, the GenoMass program could be developed to have real-time instrument control to automatically start MS/MS with found ions of interest.

Chapter 4

Investigation of Sequence Selectivity of AAAF Adduct Formation on Digestion Products of Oligonucleotides and DNA by Ion Pairing LC-MS and LC-MS/MS – A New Approach

4.1 Introduction

The significance of preserving the integrity of DNA is well established. The health of humans is heavily dependent on the correct constitution and function of these genetic codes. Yet DNA is susceptible to various attacks from a lot of species in the environment, and constantly undergoes deleterious modifications, either spontaneously or by the action of physical or chemical agents which may lead to its malfunction. Compounds that can interact with DNA include some carcinogens and oxidants that are covalently bound to DNA and form so-called DNA adducts. Adduct formation may cause major distortions in DNA, and even form crosslinks between two strands resulting in mutagenesis, inhibition of DNA synthesis, or termination of transcription. Thus, if the adducted sites are not repaired properly it is reasoned that DNA adduct formation may ultimately lead to cancer. Consequently, the modification of DNA by carcinogens is regarded as a critical step in cancer development and DNA adducts are viewed as potential biological markers to assess genetic damage and to determine to a certain degree the risk for cancer (Poirier and Beland, 1992).

There is increasing interest in cancer research to determine the structure of and to elucidate the mechanisms by which these DNA modifications can cause cancer (Basu and Essigmann, 1988). Various techniques are employed in the analysis of DNA adducts, among them ^{32}P-postlabeling, immunoassays, fluorescence spectroscopy and mass spectrometry (De Kok et al., 2002; Farmer et al., 2005; Singh et al., 2006). ^{32}P-postlabeling has been the most widely used technique owing to its high sensitivity; however, the technique is labor intensive, has to deal with radioactive materials and cannot provide structural information. For these and other reasons, the ^{32}P-postlabeling assays are gradually being replaced by mass spectrometric techniques which are noted for their constantly improving sensitivity and mass accuracy and, of course, for their ability for structure elucidation of unknown DNA adducts by MS/MS. When coupled to separation techniques such as HPLC and CE, MS and MS/MS are particularly advantageous in analyzing oligonucleotides, DNA and their adducts in complex mixtures as reported by several research groups (Andrews et al., 2004; Barry et al., 1996a, 1996b; Chen et al., 2004; Chou et al., 2003; Glover et al., 1998; Harsch and Vouros, 1998; Harsch et al, 1998, 2000; Janning et al., 1994; Liao et al., 2007; Marzilli et al., 1998; Schrader and Linscheid, 1995, 1997; Song et al., 2005; Tretyakova et al., 2002; Van den Driessche et al., 2004a; Wang et al., 1999; Xiong et al., 2007).

In most DNA and DNA adduct studies by LC-MS, the procedure typically starts with enzymatic digestion of oligonucleotides, DNA and their adducts to individual mononucleotides or mononucleosides. However, a disadvantage of this approach is the loss of sequence information as carcinogenic compounds are known to react preferentially with DNA base sequences that are called chemical hotspots. Conversely, the mutations induced by carcinogens are not distributed randomly along the DNA molecule; rather, they occur at a limited number of loci that are called mutational hotspots. The chemical hotspots do not necessarily correspond to the mutational hotspot and the relationship between the two is not yet established. It is however known that different types of chemical lesions can have different biological consequences. Some lesions lead to a significant disruption of the double helix around the adduct whereas others can be accommodated within the minor groove of the DNA with much less disruption of the local helical region. The mutation depends largely on the repair by the cellular repair system. The lesions which are not repaired efficiently remain in place long enough to exert significant biological effects and the rate of repair can differ greatly on different modification sites. It remains a question whether the mutational hotspots are dictated by the chemical hotspots or the modification reaction. It is however reasonable to hypothesize that the specific biological effects may depend on preferential reaction at certain genomic locations.

In line with the above considerations, in order to better understand the relationship between chemoselectivity and mutagenic activities, it is desirable to digest adducted DNA to oligonucleotides that are of sufficient length to provide information of the selectivity of adduction within a relevant DNA genetic sequence and which can be separable by HPLC and are amenable to MS/MS analysis. Short lengths are of limited value in constructing the correct sequences of DNA while larger (especially isomeric) pieces become progressively more difficult if not impossible to separate by HPLC or be accessed by the average MS analyzer. As a consequence, enzyme selection and optimization of enzymatic conditions are crucial for the study of oligonucleotide adducts by MS. The long established enzymes used to digest oligonucleotides, DNA and their adducts to nucleoside and nucleotide level are DNase I and Nuclease P1. For digesting into larger fragments, there are two types of enzymes: the specific and non-specific endonucleases. The specific enzymes have to function with a specific sequence but when working with an unknown sequence such as DNA, a non-specific, random endonuclease is needed. There are few candidates within the latter category that possess the required properties, and their enzymatic characteristics are not fully defined. Among them, the benzonase/alkaline phosphatase pair is considered a suitable enzyme system for digesting both single stranded (ss) and double stranded (ds) DNA and RNA into oligonucleotides of the type n-nucleotide-(n-1)-phosphate with chain length from 2-8 bases. Benzonase was produced and purified from *E. coli* strain W3110, a mutant of strain K12, as a precursor of 266 amino acids that is proteolytically processed upon secretion to yield a non-covalent dimer of 245 amino acids. Alkaline phosphatase removes free 5'-phosphate groups (Schrader and Linscheid, 1997). Another candidate is the uvrABC protein, yet it is hard to obtain this DNA repair protein that functions also as an enzyme to cleave DNA.

Linscheid and coworkers were the first to use benzonase to digest DNA to 2-8 mers for capillary zone electrophoresis mass spectrometry (CZE/MS) analysis (Janning et al., 1994). In our lab, Andrews et al. (2004) used the benzonase/alkaline phosphatase pair to digest DNA samples and analyzed the trimer products by LC-MS and LC-MS/MS. Their profiles were compared to those of digested model oligomer standards. We have most recently investigated the enzymatic behavior of benzonase/alkaline phosphatase in detail by mass spectrometry and found that the pair is in fact more non-specific in its cleavage of native DNA than had been found out by other methods (Liao et al., 2007). Moreover, this enzyme can digest DNA not only up to 8-mers as reported before but even up to 10-mers which can enable us to analyze digestion products/fragments of longer length.

Sequence selective reactions with DNA have been observed in the case of several compounds, including bleomycin, AAAF, mitomycin, benzo(a)pyrene, aflatoxins, cis-dichloridiamine platinum(II) and chloroethyl-nitrosoureas (Belguise-Valladier and Fuchs, 1991; Colgrave et al., 2003; Hartley et al., 1992; Koehl et al., 1989; Kohn et al., 1987; Mattes et al., 1986). Efforts have been made to identify and characterize the hotspots in DNA by LC-MS means on many different compounds. The effects of nucleotide sequence context on the covalent reactions of these compounds have been described and recently reviewed (Rogozin and Pavlov, 2003; Rogozin et al., 2005). Among these compounds, AAAF, a potent carcinogen, presents itself as an excellent model molecule for establishing analytical methodologies in this area. This molecule reacts at a high yield specifically with guanine residues in DNA (Fuchs, 1984), forming a covalent bond predominantly between the 2-amino group of AAF and C8 and N^2 sites of the deoxynucleotide guanine (dG) in DNA to produce two modification products (Kriek et al., 1967). The major product, N-acetyl-N-(guan-8-yl)-2-aminofluorene (C8-AAF-dGuo), formed by reacting of AAF at the C8 position of guanine, the AAF moiety is then thought to rotate into the base stack while the guanine rotates out. The structures of these adducts are shown in **Figure 4.1**.

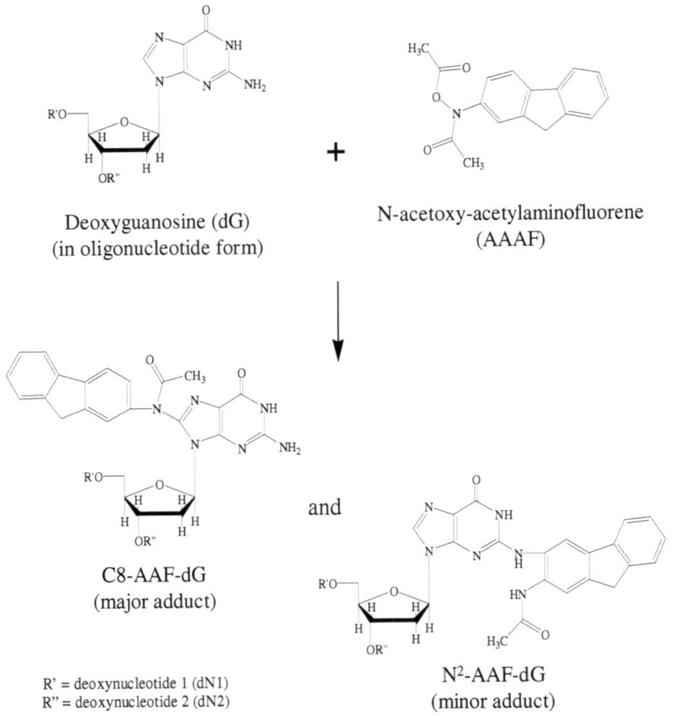

Figure 4.1. In vitro modification procedure for *N*-acetoxy-acetylaminofluroene and deoxyguanosine residues in DNA and synthetic oligonucleotides. MW of AAF = 221.09 da.

AAAF can serve as a valuable model system for other C8-binding mutagens, such as dietary carcinogens and nitroaromatic compounds which also exert their action mostly through specific binding to the C-8 position of guanine (Hoffmann and Fuchs, 1997). In spite of relatively uniform distribution of adducts in DNA, analysis of mutation spectra has indicated strong hotspots at only a limited number of sites for the induction of mutation by AAF (Seeberg and Fuchs, 1990). However, there seems to be no direct correlation between the DNA modification spectrum and the mutation spectrum. There is no preferential modification reaction at the guanine residue that is located in hotspots. The mutational specificity is not dictated by the specificity of the modification reaction but, rather, is the consequence of the premutagenic lesion. When a premutagenic lesion occurs within a mutation-prone sequence it can be converted efficiently into a mutation. When a sequence is said to be mutation-prone it probably corresponds to a particular structure that is induced within the sequence by the adduct(s) (Fuchs, 1984). Yet no simple "rules" governing the reactivity of a given G in relation to its nearest neighbors can be seen. The binding of AAF to guanine is influenced by the base sequence of the DNA but, as stated above, the relationship is complicated and the statistics are not established.

Normally, molecular mass determination alone is insufficient for unambiguous identification of oligonucleotide, DNA and DNA adducts, since only nucleobase composition is determined. Base sequence information is required for accurate structural elucidation and can be obtained using collision-induced dissociation (CID), which permits unambiguous assignment of the major sites of modification of DNA adducts within the target sequences and the complementary strand (McLuckey et al., 1992). However, in a very recent publication, Gao, Chiarelli et al used on-line HPLC-ESI-MS for characterization of oligonucleotide adducts produced from the enzymatic digestion of model AF-modified dodecamer known sequence (Gao et al., 2008). The authors sequenced six isomeric AF-modified isomers of the 12-mer 5'-CTCGGCGCCATC-3' NarI sequence by enzymatic digestion with 3'- and 5'-exonucleases and LC-ESI/MS analysis. In their study, MW was measured to assign the composition of the oligonucleotides. The assignment was based on the accuracy and precision of the measurements and, more importantly, the minimum mass difference one may expect between oligomer digestion fragments. Ions generated by 3'PDE (snake venom phosphodiesterase) digestion are referred as B_x-type ions and those generated by the 5'-PDE (bovine spleen phosphodiesterase) were referred to as Y_y-type ions, where the subscripts x and y specify the number of nucleotides associated with the respective ion B and Y. Sequence-specific fragments in these digestion analyses would permit the sequence assignment of AF-modified oligonucleotides to establish the modification sites. As in their studies the modified oligomers all have the same base sequence, the key to differentiating the positional isomers is simply to identify the base compositional oligomer digest ions that are specific for a particular sequential isomer. Their work indicated that this methodology is useful for sequencing modified oligonucleotides with single or multiple adduction sites to serve as templates for site-specific mutagenesis studies.

In this report we present the application of the LC-ESI-MS technique to the analysis of n-mer (n=4-9) deoxynucleotide adducts generated by the in vitro reaction of AAAF with DNA. The aim of this study was to investigate the sequence selectivity of reactions of AAAF with model oligonucleotides and DNA, and to develop LC-MS or LC-MS/MS methodology for the analysis of complex mixtures of oligonucleotide adducts. The study is meant to establish a sound method to assess the degree of the DNA-AAF adduct formation on different sites along the DNA chain and their possible relationship (if there is indeed any) to the carcinogenic mutation. AAAF is known to induce mutation within the recognition sequence for the NarI restriction site, -$^1G^2GC^3GCC$- (Seeberg and Fuchs, 1990). One dG residue – ^3G is a mutational hotspot while $^1G^2G$ give no rise to mutation. The reactivity of AAAF with different G sites was examined by studying several single stranded and double stranded model oligonucleotides that were designed around the NarI sequence. Then the same approach was applied to ctDNA, where we used again the automated data analysis tool GenoMass software to find AAF modified digestion products and to compare their relative abundances semi-quantitatively.

4.2 Experimental Section

4.2.1 Chemicals

The ss 4-mer oligonucleotide standards and ss model NarI oligonucleotides were obtained from Sigma Genosys (The Woodlands, TX). Calf thymus DNA (ctDNA) was purchased from Sigma-Aldrich (St. Louis, MO). The duplex oligonucleotides were purchased from Integrated DNA Technologies, Inc. (IDT) (Coralville, IA). Benzonase was purchased from Novagen (Madison, WI). Alkaline phosphatase was purchased from Sigma-Aldrich (St. Louis, MO). AAAF was purchased from NCI Chemical Carcinogen Repository (Rockville, Maryland). *AAAF is carcinogenic and should be handled carefully.* Citrate buffer was obtained from Sigma-Aldrich (St. Louis, MO). Tris/$MgCl_2$ buffer was obtained from Schwarz/Mann Biotech (Cleveland, OH). Triethylammonium bicarbonate (TEAB) was obtained from Sigma-Aldrich (St. Louis, MO). All solvents were HPLC grade unless specified otherwise. Acetonitrile, methanol, and ethyl acetate were obtained from Mallinckrodt Baker (Paris, Kentucky). Water was from EMD Chemical (Gibbstown, NJ). Oasis SPE cartridges were obtained from Waters (Milford, MA).

4.2.2 Methodology

4.2.2.1 Sample Preparation

Stock Solution Preparation

The tubes containing **x** nmol 4-mer standards, ss and ds oligonucleotide standards were centrifuged prior to opening. Then **x** μL DI water was added to the 4-mer oligonucleotide standards and the ss NarI 12-mer, NarI 20-mer, and NarI 21-mer oligonucleotides respectively to create stock solutions of concentration 1 nmol/μL for each sample. The **x** nmol ds NarI 12-mer, NarI 20-mer, NarI 21-mer were resuspended with **x** μL RNase-Free Duplex Buffer to make a 100 μM solution. These mixtures were all heated to 94 ^{o}C for 2 minutes, removed from heat and allowed to cool slowly to room temperature. This resulted in stable, double stranded form of oligomers and they were stored at –20 ^{o}C. 1 mg ctDNA was weighed out and 1 mL 10 mM citrate buffer (pH = 6) was added to a concentration of 0.1 nmol/μL. All samples were then vortexed down.

AAAF Modification

35.5 μL of 1 nmol/μL 4-mer oligonucleotide standards, ss and ds NarI 12-mer, NarI 20-mer, and NarI 21-mer oligonucleotides were centrifuged down. 355 μg ctDNA was added to 355 μL of 10 mM citrate buffer (pH = 6) to a concentration of 0.1 nmol/μL DNA. The above oligonucleotide solutions and DNA were mixed with the AAAF stock solution of 3.55 nmol/μL in acetonitrile in 1:1 molar equivalence. The reaction was allowed to progress at room temperature. The mixtures were then incubated on a heat block at 37 ^{o}C for 18 hours. The modified 4-mer standards were reconstituted in 100 μL deionized water and further diluted 1:100 for LC-MS analysis.

Enzymatic Digestion

The modified DNA was extracted 3 times with 1 mL ethyl acetate to remove the unreacted AAAF. The aqueous fractions were then dried down using a Speedvac from Savant (Pleasanton, CA) for 2 h at medium heat. The modified oligomers and DNA were then reconstituted in 200 μL of 5 mM Tris/10 mM $MgCl_2$ buffer (pH = 8). After being vortexed down, 5 μL benzonase and 5 μL alkaline phosphatase were added to the oligomers and DNA in buffer solution. The mixtures were vortexed well and incubated for 24 h at 37 ^{o}C.

Solid Phase Extraction

Upon completion of the digestion, the mixture then went through solid phase extraction (SPE) to concentrate the DNA adducts and remove the unadducted DNA fragments. SPE was performed by using a vacuum manifold fitted with disposable Teflon needles to prevent sample cross-contamination. Unmodified oligomers were washed out while modified oligomers were eluted. The optimized SPE conditions were as follows: (1) Conditioning with 1 ml of methanol. (2) Conditioning with 1 ml of Water. (3) Load the samples. (4) Wash with 1 ml 5% (v/v) methanol/water. (5) Elute with 75% (v/v) methanol/water. The adducts-containing fraction was dried down and reconstituted in 100 μL water for analysis.

4.2.2.2 Chromatographic Conditions

The LC system employed was an Agilent Technologies 1090 HPLC with an autosampler (Wilmington, DE). The HPLC system was controlled by HP ChemStation software running on an IBM-compatible PC. The reverse phase RP-1 PSDVB (polystyrene divinylbenzene) column (2.1 mm x 100 cm x 3.5 μm) was purchased from Hamilton Chromatography (Reno, NV). The column was kept at ambient temperature (25 °C) during analysis. The mobile phase consisted of 25 mM TEAB as A and 25 mM TEAB with 20% acetonitrile as B. The gradient went from 0%B to 100%B in 20 minutes and was held at 100%B for 5 minutes. The flow rate was 0.2 mL/min and the injection volume was 20 μL.

4.2.2.3 Mass Spectrometric Conditions

MS data were acquired on a ThermoFinnigan LCQ Classic mass spectrometer (San Jose, CA). The mass spectrometer was tuned on a series of standard oligonucleotides. The samples were analyzed by negative ion electrospray mass spectrometry, first in full scan mode, and subsequently by MS/MS. The MS conditions were as follows: spray voltage 3.8 kilovolts, capillary voltage -4 volts, capillary temperature 200 °C, sheath gas 25 arbitrary units, tube lens voltage -25 volts. MS/MS experiments were conducted at a collision width of 1.5 Dalton and relative energy of 30%. Data were acquired and processed with Navigator 1.2 software. Mass spectra were recorded in continuum mode.

4.2.2.4 Computer Software

GenoMass is a computer software that can automate the processing of data from the LC-MS and LC-MS/MS analysis of oligonucleotide, DNA and DNA adducts (Liao 2007). The program automatically calculates masses of oligomers or/and their adduct ions, and searches for them through the LC-MS data, then outputs a list of found fragments, their abundance and the corresponding oligonucleotides and their isomers along with other pertinent information. The program is also capable of generating extracted ion chromatograms (EIC) based on the found masses on a click of the mouse. It can search masses within a defined retention window and can also perform background subtraction. The software is designed to eliminate the otherwise extraneous manual labor demanded by large amounts of data produced by MS analysis on DNA and DNA adducts. Full details of the software have been described in a recent publication (Liao, Shen, Vouros 2009) and in chapter 3 of this book.

4.3 Results and Discussion

In a recent publication in which we investigated the enzymatic cleavage of oligonucleotides and DNA by benzonase/alkaline phosphatase, we reported that the digestion process showed a strong preference for the production of 4-mer fragments (Liao et al., 2007). Accordingly, in the present study which was aimed at the evaluation of the sequence selectivity of AAAF in its reaction with oligonucleotides and DNA, we focused our attention initially on the search for modified 4-mers detected in the LC-MS analysis of the digest. In line with this strategy, we first sought to establish the optimal conditions for LC-MS analysis of AAF-4-mer adducts produced by reacting standards of twenty tetramers with AAAF. The structures of the 4-mer standards selected for the study are shown in **Table 4.1**. The selection of these 4-mer standards was based on the following criteria: There are a total 35 possible combinations of 4 bases into 4-mers of the same mass; each combination can have one to 24 isomers; among these masses, 20 are the ones that contain at least one G. We thus chose one isomer of each mass as a standard.

Table 4.1. 4-mer standards studied in the investigation of the selectivity of DNA-adduct formation.

Oligo Name	Sequence	M	$[M - H]^-$	$[M - 2H]^{2-}/2$	$[M\text{-}AAF - H]^-$	$[M\text{-}AAF - 2H]^{2-}/2$
Oligo tetramer 1	GCCC	1134.8	1133.8	566.4	1354.8	676.9
Oligo tetramer 2	TGCC	1149.8	1148.8	573.9	1369.8	684.4
Oligo tetramer 3	AGCC	1158.8	1157.8	578.4	1378.8	688.9
Oligo tetramer 4	TTGC	1164.8	1163.8	581.4	1384.8	691.9
Oligo tetramer 5	ATGC	1173.8	1172.8	585.9	1393.8	696.4
Oligo tetramer 6	GGCC	1174.8	1173.8	586.4	1394.8	696.9
Oligo tetramer 7	TTTG	1179.8	1178.8	588.9	1399.8	699.4
Oligo tetramer 8	AAGC	1182.8	1181.8	590.4	1402.8	700.9
Oligo tetramer 9	ATTG	1188.8	1187.8	593.4	1408.8	703.9
Oligo tetramer 10	TGGC	1189.8	1188.8	593.9	1409.8	704.4
Oligo tetramer 11	AATG	1197.9	1196.9	598.0	1417.9	708.5
Oligo tetramer 12	AGGC	1198.8	1197.8	598.4	1418.8	708.9
Oligo tetramer 13	TTGG	1204.8	1203.8	601.4	1424.8	711.9
Oligo tetramer 14	AAAG	1206.9	1205.9	602.5	1426.9	713.0
Oligo tetramer 15	ATGG	1213.9	1212.9	606.0	1433.9	716.5
Oligo tetramer 16	GGGC	1214.8	1213.8	606.4	1434.8	716.9
Oligo tetramer 17	AAGG	1222.9	1221.9	610.5	1442.9	721.0
Oligo tetramer 18	TGGG	1229.9	1228.9	614.0	1449.9	724.5
Oligo tetramer 19	AGGG	1238.9	1237.9	618.5	1458.9	729.0
Oligo tetramer 20	GGGG	1254.9	1253.9	626.5	1474.9	737.0

These standards were also used to establish a semi-quantitative base for analysis of oligonucleotide adducts and, more specifically a comparison of the relative responses of modified isomeric compounds. For comparison, data from UV (ultraviolet) detection was also analyzed. The UV data are in general in agreement with the MS data (**Figure 4.2**).

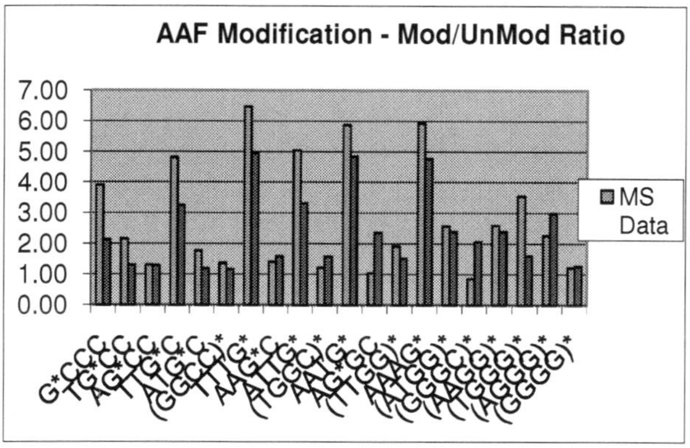

Figure 4.2. The comparison of Mod/UnMod Ratio of AAF modification on 4-mers standards by MS and UV.

Since it is generally known that UV is a more direct quantitative method than MS, UV data were obtained and correlated with MS data in order to support the semi-quantitative approach taken in the MS analysis of the4-mer standards. As shown in **Figure 4.2**, the ratios of the UV and MS data on both the unmodified and modified 4-mer products were in agreement with each other. Thus it was reasonable to rely on the MS data to establish proportions of isomeric products at least in a semi-quantitative context.

The conditions used for analysis of the 4-mers were subsequently applied to the analysis of digests of progressively larger AAAF-modified oligonucleotides and finally ctDNA whose structure is unknown. The structures of the oligonucleotides selected for the study are shown in **Table 4.2**. The 12-mer NarI gene sequence shown in blue and which contains a chemical hot spot (shown in red) for AAAF has been incorporated in all of the oligomers.

Table 4.2. Model oligonucleotide sequences studied in the investigation of the selectivity of DNA-adduct formation.

Oligonucleotide ID	**Sequence**	**MW** (mono) (da)
NarI 12-mer-1	5'-ACCGGCGCCACA-3'	3598.6
NarI 12-mer-2	3'-TGGCCGCGGTGT-5'	3691.6
NarI 12-mer Duplex	5'-ACCGGCGCCACA 3' 3'-TGGCCGCGGTGT-5'	7290.3
NarI 20-mer	5'-CCCAACACCGGCGCCACAGA-3'	6023.1
NarI 21-mer-1	5'-GCCCAACACCGGCGCCACAGA-3'	6352.1
NarI 21-mer-2	3'-TAGGGTTGTGGCCGCGGTGTA-5	6546.1
NarI 21-mer Duplex	5'-ATCCCAACACCGGCGCCACAT-3' 3'-TAGGGTTGTGGCCGCGGTGTA-5	12898.2
Calf thymus DNA	unknown	

4.3.1 LC-MS Analysis of Standard 4-mer Adducts

The standard oligonucleotide tetramers were subjected to the same AAAF modification reaction and cleanup procedures as those for model oligonucleotides and the ctDNA and the reaction mixtures analyzed by LC-MS/MS. In an effort to achieve optimal chromatographic efficiency and response, several systems and ion-pairing mobile phase combinations were evaluated, among them: HFIP, ammonium acetate, triethyammonium acetate (TEAA) and triethylammonium bicarbonate (TEAB) systems. The HFIP system commonly used in oligonucleotide separations by Apffel et al. (1997a, 1997b) presented a problem due to the persistence of a m/z 213 ion and other background which resulted in lowered sensitivity of the instrument in the negative ionization mode. TEAB in combination with a PSDVB column provided the best separation and highest sensitivity of 4-mers and their modified isomers and thus was used for all online chromatographic separations.

Triplicate samples were processed together with the blank controls under the same LC-ESI-MS conditions to ensure precision. The chromatographic analysis of the adducts was also enhanced by solid phase extraction (SPE) of the reaction mixture, which removed the bulk of the unmodified oligonucleotides and provided maximum adducts enrichment in the reconstituted sample. Unmodified 4-mers eluted first and the more hydrophobic modified 4-mers eluted later, with the smaller oligomers eluting earlier than larger oligomers. Significantly, isomeric 4-mers that differed only in the location of the covalent modification within the oligonucleotide sequences were well separated. MS and MS/MS using an LCQ ion trap MS provided molecular weight and fragmentation information for identification of unknowns (Liao et al., 2007). The 4-mers were mainly detected as singly charged ions, although some doubly charged ions were still present. All extracted ion profiles were based on the mass of singly charged DNA-adduct species. The detection limit of the system was found to be approximately 0.2 pmol, or 0.27 ng for the 4-mer standards, 1 pmol, or 1.35 ng for mixture, with an injection volume of 20 μL. The experiments were first performed on the known oligonucleotide standards at a concentration of 0.1 μM. An example of the separation of unmodified and modified 4-mers is shown in **Figure 4.3**. Unmodified 4-mers eluted first and the modified 4-mers eluted later as AAF is hydrophobic. The 4-mers are fairly non-polar too.

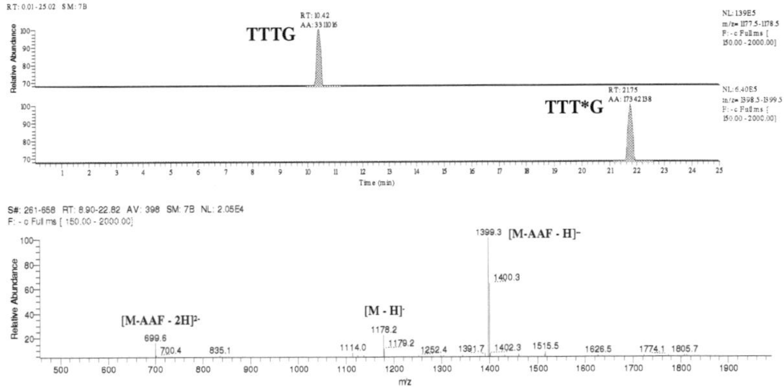

Figure 4.3. The separation of unmodified and modified 4-mer standards. * denotes the modification site.

All the retention times of unmodified 4-mer standards fall into the window between 7-11 min, and the modified 4-mer standards fall in the window of 16-22 min.

Figure 4.4 is an example of an LC profile (TIC) and a spectrum of an AAF modified GGCC 4-mer standard with a M/Z of 1394. Two isomers G*GCC and GG*CC were fully resolved. It should be noted that even in cases where chromatographic peaks coeluted, MS/MS provided sufficient selectivity for separation of isobaric 4-mer adducts. The distinct features of LC-MS/MS are shown in **Figure 4.6** which compares the CID spectra of G*GCC and GG*CC.

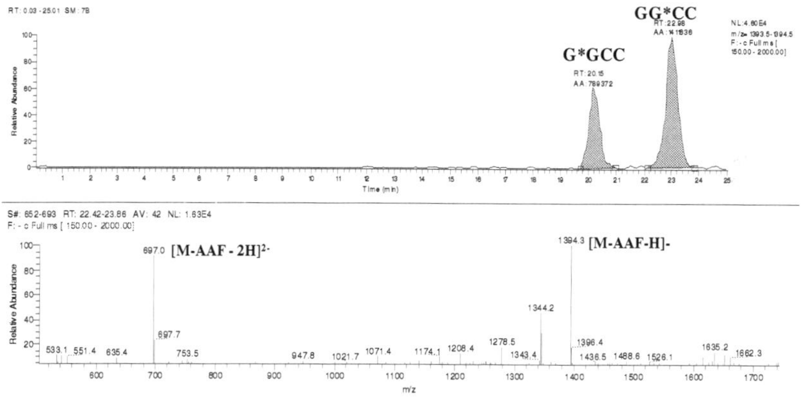

Figure 4.4. A total ion chromatogram (TIC) and a spectrum of an AAF modified GGCC 4-mer standard with a M/Z of 1394.

Figure 4.5 shows the separation of three isomers of 1449.5.

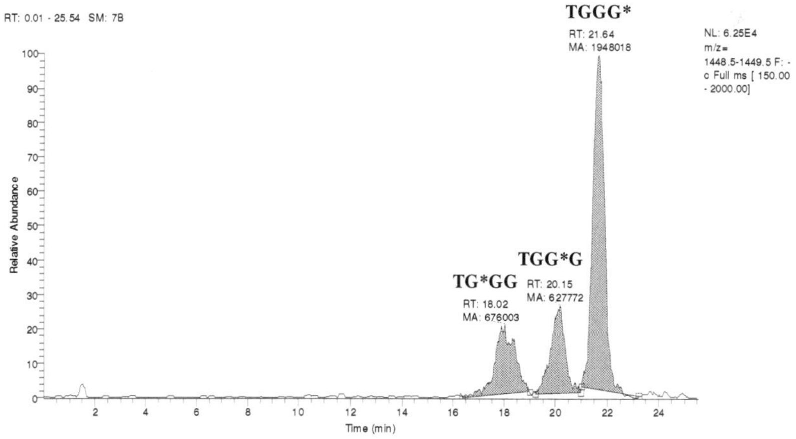

Figure 4.5. Three isomers of 1449.5 were fully resolved.

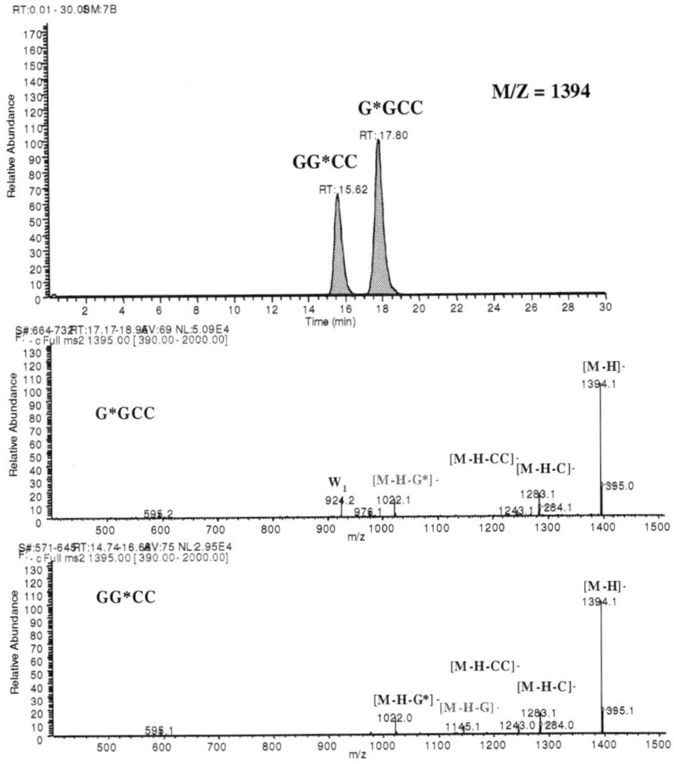

Figure 4.6. The MS/MS of [GGCC]*.

4.3.2 AAF Adducts with Oligonucleotides and DNA - Determination of the Sequence Selectivity of AAF Modification

The NarI sequence, which is known to contain a mutation hot spot at ^{3}G, was used to evaluate the potential utility of the GenoMass software for distinguishing between adduction site and mutation. This 12-mer sequence was also incorporated into the 20-mer and 21-mer oligonucleotides of known overall sequence. The NarI 12-mer, 20-mer, 21-mer were designed to contain both symmetric and asymmetric sequences. The study was conducted with both single stranded and double stranded model oligonucleotide sequences bearing in mind that ds mimics more closely the 3-D structure of the native DNA. Both ss and ds oligonucleotides went through the same AAF modification, enzymatic reaction, SPE and LC-MS analysis procedures.

It was reasoned that successful demonstration of the strategy to the analysis of digests from oligomers of known sequence would pave the way for applications to larger DNA strands. This strategy is justified by the fact that in typical studies of DNA adducts, genomic DNA is first cut into segments of varying length using restriction enzymes which could then be accessed using either the present or more advanced versions of the software as proposed later in this thesis. Since AAAF binds primarily to a G-base, a semi-quantitative assessment of the binding preference could be obtained by comparing the ion currents of modified oligomers (4-mer through 8-mer) produced in the enzymatic digestion of the model oligonucleotides. In that regard, the 4-mer standards were used to establish the relative response ratios of G-containing AAF adducts, as described earlier. The ion masses and base compositions of the digestion products were recognized by the GenoMass software and, since the sequence of the parent oligomer was known, the fragments selected to plot the reconstructed chromatograms could be directly related to individual segments that did (or did not) include the mutation hotspot ^{3}G.

AAF is a bulky adduct bound to a base instead of disrupting the sugar-phosphate backbone. In the past it was found by LC-MS that AAF mostly bound to guanine. In order to study the AAAF modification behavior on oligonucleotides and DNA, the oligonucleotide standards and DNA were reacted with AAAF under conditions identical to those used for the 4-mers, taken through enzymatic digestion and SPE steps as described in **Section 4.2.2.1** and then directly subjected to LC-MS analysis using negative ion detection. The results obtained from the digestion of the model oligomers and DNA are presented next.

4.3.2.1 Sequence Selectivity in Model Oligonucleotides

The NarI sequence 5'ACC^{1}**G**^{2}**G**C^{3}GCCACA-3' which is known to contain a mutational hotspot at ^{3}**G** was used as a model in the form of 12-mer, 20-mer and 21-mer sequences in order to address by mass spectrometric means the question of whether a mutation 'hotspot' demonstrates a high degree of adduct formation.

For all the studied oligonuleotide adducts, we evaluated different fragments ranging from 4-mers to 9-mers, and compared the abundance of adducted species containing the mutational hotspot G (denoted in RED) to the ones containing the non hotspot Gs. In this semi-quantitative approach, we were able to draw our conclusion based on the relative ratios of these species and also on the previous studies which showed us there was not a significant difference in ionization efficiency between oligonucleotide adducts. The differentiation of the isomers resolved by LC-MS could be attained by examining the internal fragments alone, and MS/MS data could provide further confirmation.

Single Stranded NarI 12-mer

Figures 4.7, **a-e** show the extracted ion chromatograms (EIC) of AAF- adducted 4-mer, 5-mer, 6-mer, 7-mer, and 8-mer detected in the LC-MS analysis of the digestion products of ss NarI 12-mer by the benzonase/alkaline phosphatase enzyme pair in an orderly cut manner. The 4-mer adducts GCG*C and G*CGC were well resolved and the two species were of about equal relative abundance. The 5-mer hotspot-containing adducts showed accumulated abundance of less than one third that of other adducts which do not contain hotspot ^{3}G. The most revealing information came from the 6-mer data, which showed the hotspot 5-mer G*CCACA, containing the hot spot G, at least one order of magnitude lower in abundance than that of the other species. The 7-mer data is less telling as the isomers of adducted ACCGGCG were not resolved. This is also the case with the 8-mer adducts, as the separation of the longer fragments is harder than the shorter ones. Nonetheless, all the above data suggested that the modification on the mutational hotspot G (^{3}G) is at least not the strongest, if not to say much lower than the modification on the other Gs (^{1}G and ^{2}G).

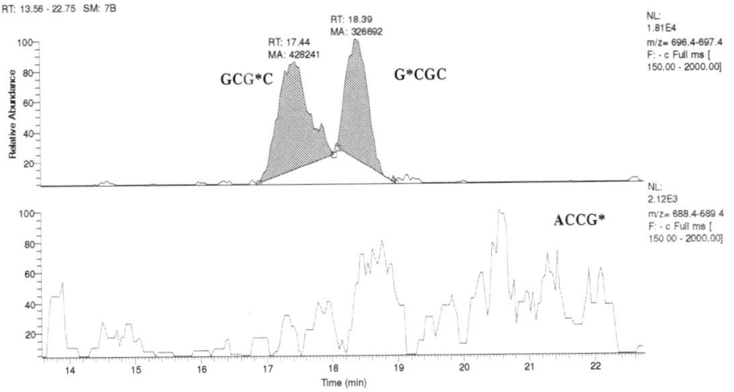

Figure 4.7a. Single stranded NarI 12-mer 5'-**ACCGGCGCCACA**-3', 4-mer adducts.

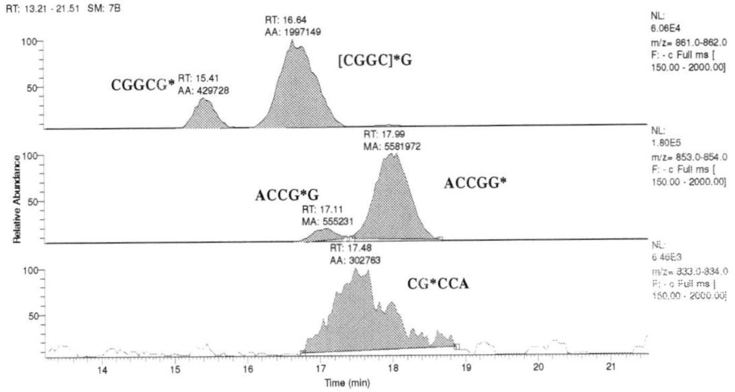

Figure 4.7b. Single stranded NarI 12-mer 5'-**ACCGGCGCCACA**-3', 5-mer adducts. **[x]** denotes the compositional isomers.

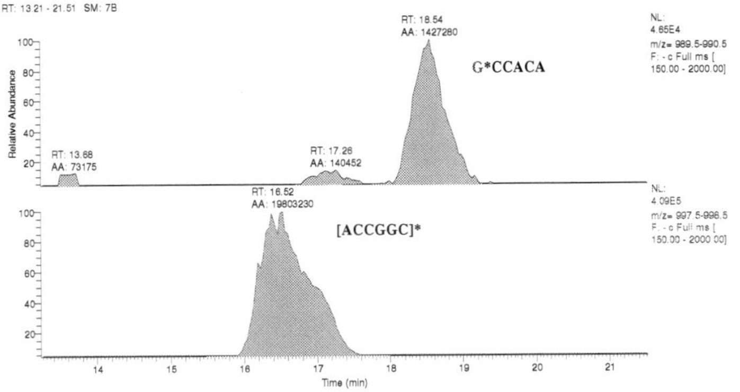

Figure 4.7c. Single stranded NarI 12-mer 5'-**ACCGGCGCCACA**-3', 6-mer adducts. **[x]** denotes the compositional isomers.

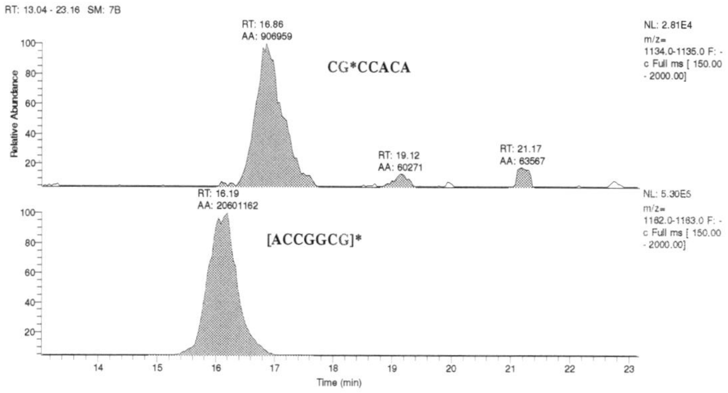

Figure 4.7d. Single stranded NarI 12-mer 5'-**ACCGGCGCCACA**-3', 7-mer adducts. [x] denotes the compositional isomers.

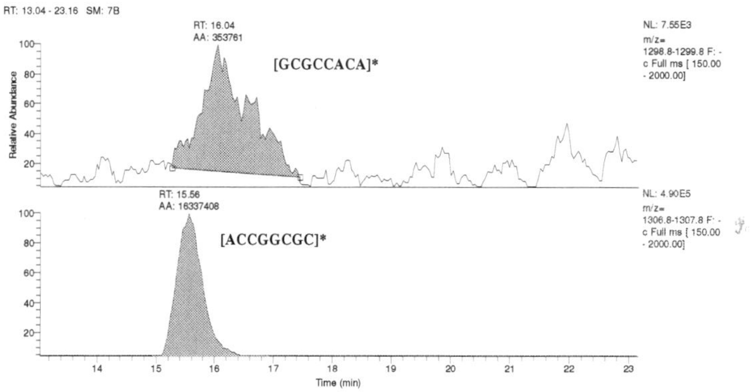

Figure 4.7e. Single stranded NarI 12-mer 5'-**ACCGGCGCCACA**-3', 8-mer adducts. [x] denotes the compositional isomers.

Single Stranded NarI 20-mer

Figures 4.8(a-c) show the EIC's of the 4-mer, 5-mer and 7-mer adducts from the enzymatic digestion of the AAAF-modified NarI 20-mer detected by LC-MS. The 4-mer adduct G*CCA was about equal in abundance to the two isomers CG*GC and CGG*C, but far less than CAG*A. The 5-mer adducts of composition GCG*CC and/or G*CGCC although they may not have separated, it is obvious that the total abundance of the two is less than a half of the adducted species ACAG*A (the adduct on ^{1}G). The 7-mer data showed the hotspot G containing fragment ACCGGCG* occurring at much lower abundance than 7-mer which do not contain the hotspot G. From these representative data, we can again at least draw the conclusion that the modification on the mutational hotspot G is not the strongest.

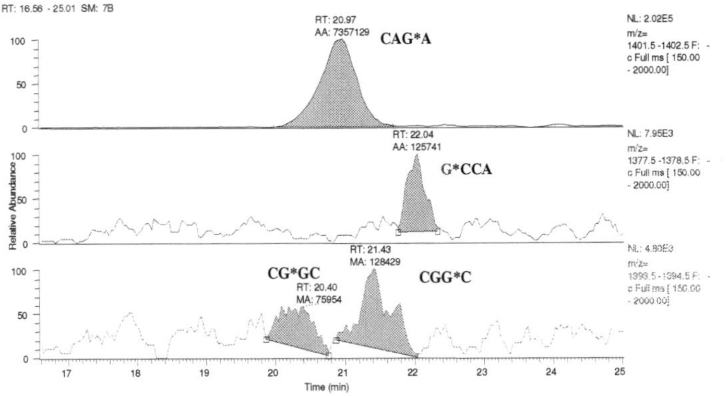

Figure 4.8a. Single stranded NarI 20-mer 5'-CCCAAC**ACCGGCGCCACA**GA-3', 4-mer adducts.

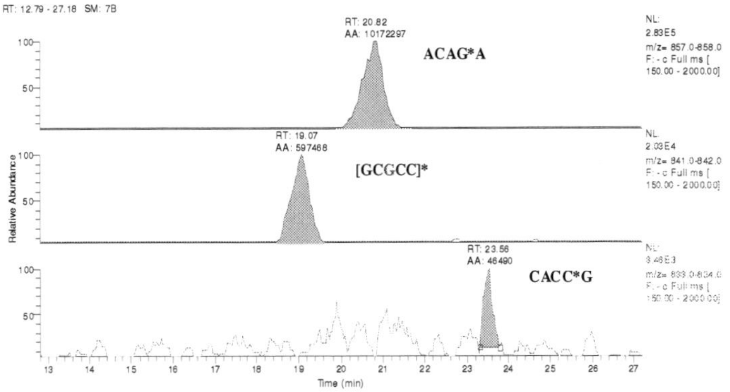

Figure 4.8b. Single stranded NarI 20-mer 5'-CCCAAC**ACCGGCGCCACA**GA-3', 5-mer adducts. **[x]** denotes the compositional isomers.

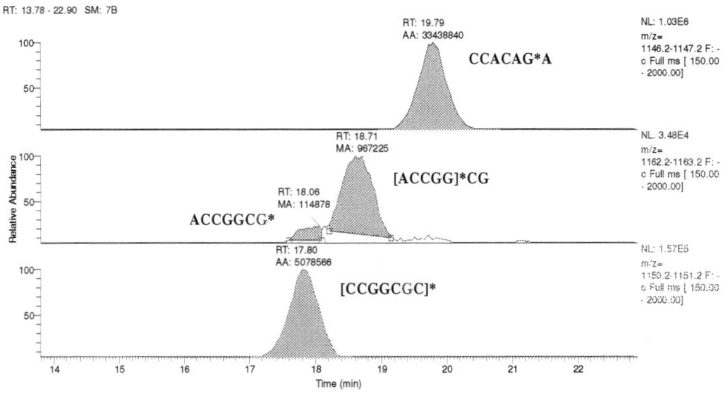

Figure 4.8c. Single stranded NarI 20-mer 5'-CCCAAC**ACCGGCGCCACA**GA-3', 7-mer adducts. **[x]** denotes the compositional isomers.

Single Stranded NarI 21-mer

In order to address a more general applicability of our approach, we examined next the NarI sequence in a 21-mer context which presents itself as an asymmetric sequence when digested into 4-mers, or 5-mers. The findings were consistent with those for the 20-mers. We have included the EICs of the 4-mer adducts for comparison with the 20-mer (**Figure 4.9**). CAG*A remained the most abundant and the non hotspot containing isomers CG*GC and CGG*C were lower in abundance while the hotspot containing G*CCA was not even detected. In another words, there are no modification products found on the site of the mutational hotspot on the ss oligonucleotide 21-mer. The conclusion was that the mutational hotspot remained the weakest site for adduct formation. In summary, all the single stranded oligonucleotides (12-mer, 20-mer, 21-mer) showed that the abundance of modified ^{3}G containing 4-mer fragments was lower than those bearing modifications in other positions.

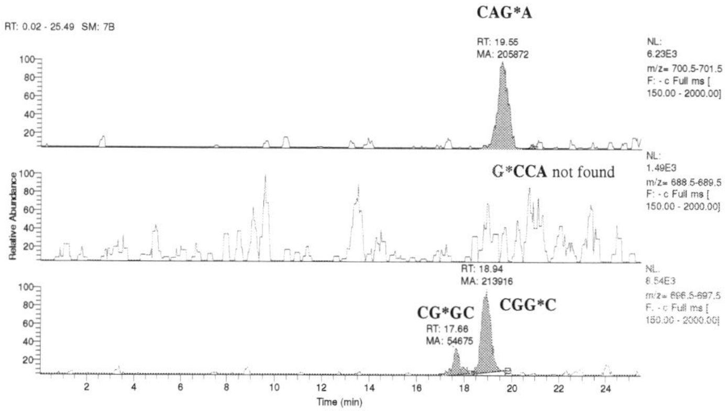

Figure 4.9. Single stranded NarI 21-mer 5'-GCCCAACACCGGCGCCACAGA-3', 4-mer adducts.

Sequence Selectivity of AAAF Adduction in Double Stranded Oligonucleotides

As stated previously, studying the behavior of the double stranded oligomer is physiologically more relevant as the ds form mimics more closely the native DNA. However, in that case, the GenoMass software has also to consider reactions (adduction) at the complementary sequence (bottom strand) in addition to those on the NarI sequence (the upper strand). The ds NarI 12-mer and 21-mer were separately reacted with AAAF and the digested mixtures analyzed by LC-MS. **Figures 4.10a** and **4.10b** show the extracted ion chromatograms of the most abundant modified 4-mers from the upper and the complementary strands, respectively formed in the reactions of the ds 12 mer. From the relative abundance of the ions, it can be deduced that the there is a preponderance of modified 4-mers generated from the bottom strand than from the top strand. This may be attributed to a larger presence of Gs in the former. To prove this point, we constructed single strands with the same sequence as the bottom strand of this oligo 12-mer and 21-mer duplex. The results showed a similar trend. This suggests that the degree of modification has more to do with the base composition than with whether it is upper or complementary strand.

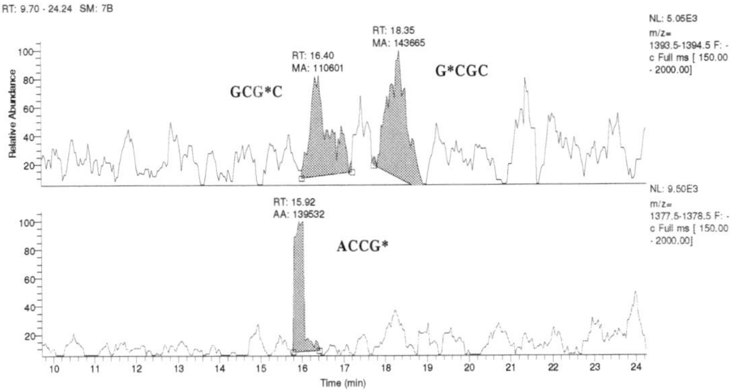

Figure 4.10a. NarI 12-mer duplex 5'-**ACCGGCGCCACA**-3',
3'-TGGCCGCGGTGT-5',
EICs of 4-mer adducts from upper strand.

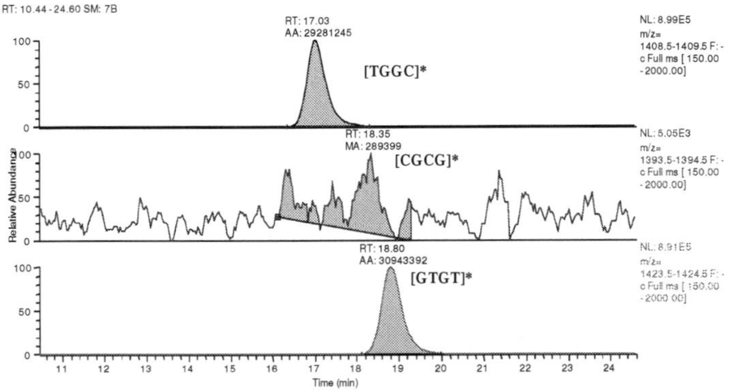

Figure 4.10b. NarI 12-mer duplex 5'-**ACCGGCGCCACA**-3'
3'-TGGCCGCGGTGT-5'
EICs of 4-mer adducts from bottom strand. [**x**] denotes the compositional isomers.

The same experiment was repeated with the double stranded NarI 21-mer. **Figures 4.11(a-f)** show the EICs of AAF-adducted 4-mer, 5-mer, 6-mer, 7-mer, and 8-mer digestion products of the upper strand of the ds NarI 21-mer produced by the benzonase/alkaline phosphatase enzyme pair in an orderly cut manner. The 4-mer hotspot-containing adduct CG*CC showed about one order of magnitude lower abundance than the adducts which do not contain hotspot G. The 5-mer data also showed the hotspot containing adduct CGGCG* was lower in abundance. The 6-mer data are not informative as the adducted CCGGCG isomers were not resolved. However, it is very clear from the 7-mer adducts data that the hotspot G containing G*CCACAT was not present while there were many other non hotspot containing adduct species. The 8-mer data was not included here as the separation was not good enough to make a sound comparison. Yet the 8-mer data demonstrated among the three AAF adducted isomers, the hotspot containing CCGGCG*CCA was not the most abundant. All the above data supported the conclusion that the modification on the mutational hotspot G (^{3}G) is lower than that on the other Gs (^{1}G and ^{2}G). The same inference was drawn from the experiments with the single stranded oligonucleotides.

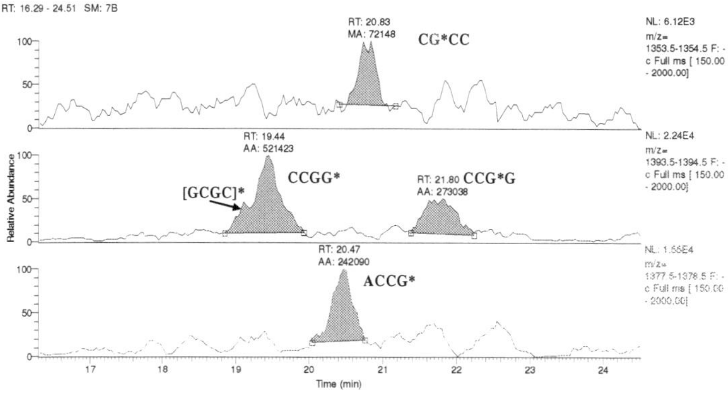

Figure 4.11a. NarI 21-mer duplex 5'-ATCCCAAC**ACCGGCGCCACA**T-3'
3'-TAGGGTTGTGGCCGCGGTGTA-5',
EICs of 4-mer adducts from upper strand. [**x**] denotes the compositional isomers.

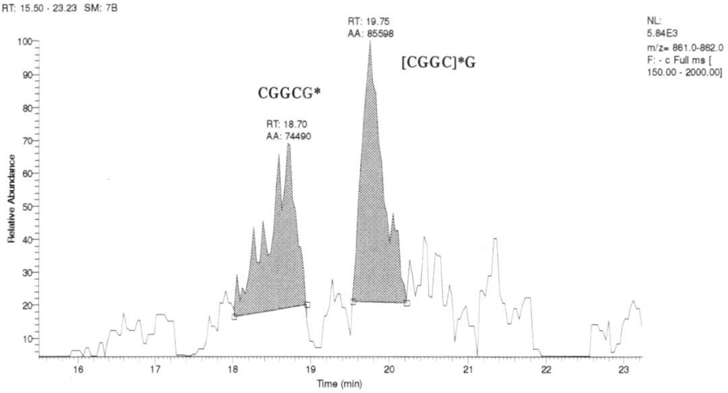

Figure 4.11b. NarI 21-mer duplex 5'-ATCCCAAC**ACCGGCGCCACA**T-3'
3'-TAGGGTTGTGGCCGCGGTGTA-5',
EICs of 5-mer adducts from upper strand. [**x**] denotes the compositional isomers.

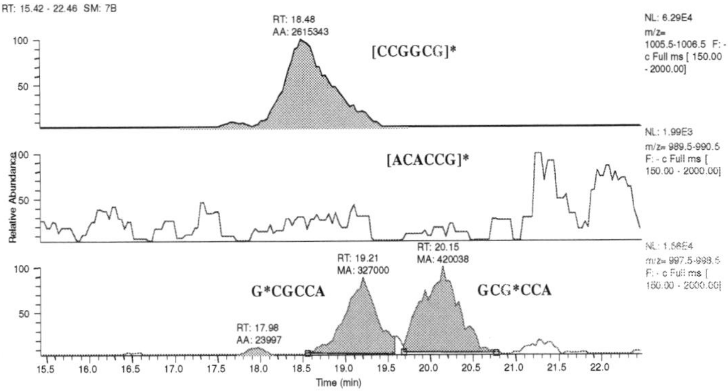

Figure 4.11c. NarI 21-mer duplex 5'-ATCCCAAC**ACCGGCGCCACA**T-3'
3'-TAGGGTTGTGGCCGCGGTGTA-5',
EICs of 6-mer adducts from upper strand. **[x]** denotes the compositional isomers.

[CCGGCG]* = **CCG*GCG + CCGG*CG + CCGGCG***

As shown in **Figure 4.11c** (lowest panel), two 6-mer adducts were well resolved chromatographically while no CID data were obtained, the positions of AAF adduction, nonetheless ESI alone was able to establish their identities. In this case, the ESI process itself induced sufficient fragmentation to provide evidence for the location of the adduct sites, as shown in **Figure 4.11d**.

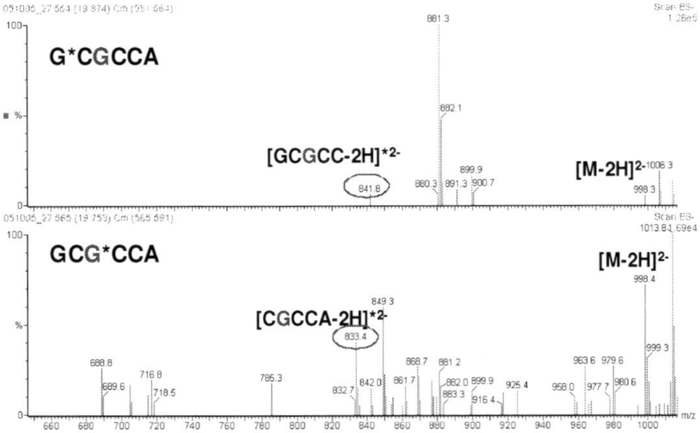

Figure 4.11d. MS spectra of **G*CGCCA** and **GCG*CCA**. **[x]** denotes the compositional isomers.

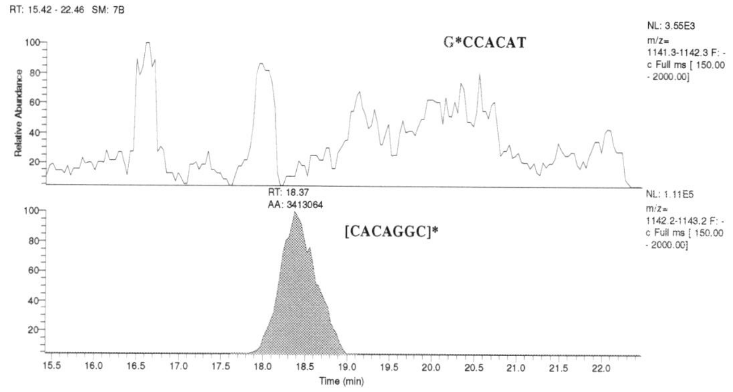

Figure 4.11e. NarI 21-mer duplex 5'-ATCCCAAC**ACCGGCGCCACA**T-3'
3'-TAGGGTTGTGGCCGCGGTGTA-5',
EICs of 7-mer adducts from upper strand. [**x**] denotes the compositional isomers.

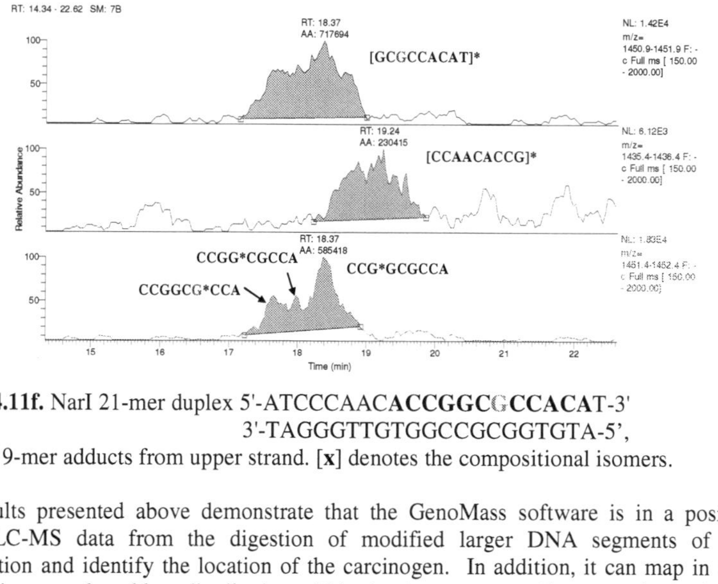

Figure 4.11f. NarI 21-mer duplex 5'-ATCCCAAC**ACCGGCGCCACA**T-3'
3'-TAGGGTTGTGGCCGCGGTGTA-5',
EICs of 9-mer adducts from upper strand. [**x**] denotes the compositional isomers.

The results presented above demonstrate that the GenoMass software is in a position to screen LC-MS data from the digestion of modified larger DNA segments of known composition and identify the location of the carcinogen. In addition, it can map in a semi-quantitative way the adduct distribution within the the sequence. For example, in the ase of the NarI gene sequence examined here it was possible to establish the binding selectivity of adduction at the three different G sites as summarized on **Table 4.3.** The data pointed to stronger bonding to AAF at ^{1}G and ^{2}G than ^{3}**G**, which is the mutation hotspot. Thus, the mutation hotspot ^{3}**G**, appears not to be the most strongly bonded to AAF by the mass spectrometric approach.

Table 4.3. Comparison of the AAF modification on different adduct bonding sites on ss and ds NarI sequences. The mutational hotspot G is in red.

Oligonucleotide ID	Sequence Selectivity	Indication
NarI 12-mer-1	↓↓ ↓ ↓ ↓ 5'-**ACCGGCGCCACA**-3'	Modification on G not the strongest.
NarI 12-mer Duplex	↓↓ ↓ ↓ ↓ 5'-**ACCGGCGCCACA**-3' 3'-TGGCCGCGGTGT-5' ↑↑ ↑ ↑ ↑	Modification on G not the strongest.
NarI 20-mer	↓↓ ↓ ↓ 5'-CCCAAC**ACCGGCGCCACA**GA-3'	Modification on G lower.
NarI 21-mer-1	↓ ↓↓ ↓ ↓ 5'-GCCCAAC**ACCGGCGCCACA**GA-3'	Modification on G lower.
NarI 21-mer Duplex	↓↓ ↓ 5'-ATCCCAAC**ACCGGCGCCACA**T-3' 3'-TAGGGTTGTGGCCGCGGTGTA-5' ↑↑↑ ↑ ↑↑ ↑ ↑↑ ↑	Modification on G lower.

4.3.2.2 Calf Thymus DNA

Following the successful demonstration of the use of the GenoMass software for analysis of enzymatic digestion products from small DNA segments of known sequence, we applied the same pseudo-combinatorial approach strategy to the study of ctDNA, whose sequence is unknown. While it was not anticipated that the study would yield definitive data with regard to sequence selectivity, it was of interest to see if any specific trends could be observed with regard to adduct formation.

In the specific study of calf thymus DNA, the adduct yield from the reaction with AAAF appeared to be too low for the chromatographic conditions employed with the LCQ ion trap. Because of the increased sensitivity requirements, we switched to use of a monolithic column (75 μm i.d, 360 μm o.d, 15 cm) with a nanoLC (Eksigent ExpressLC-100) – μESI ionization system interfaced to a Waters Q-Tof micro MS. The system was operated at a flow rate of 200 nL/min, however all other conditions (solvent composition and ion-pairing reagent) remained the same. The LC system was controlled by the MassLynx 4.1 software. The monolithic column was made in house with a PSDVB stationary phase. The gradient initially stayed at 98%A, 25 mM TEAB for 3 minutes, then went from 98%A to 98%B, 25 mM TEAB with 20% acetonitrile in 40 minutes. Then the gradient was held at 98%B for 20 minutes. The MS conditions were: sample cone voltage at -55 volts, capillary voltage at 1500 volts, extraction cone voltage at 4.5 volts, cone gas at 11 L/hr, desolvation gas at 15 L/hr, source temperature at 100 °C.

The system achieved a detection limit of 1 fmol or 1.36 pg on 4-mer standards, with an injection volume of 300 nL. The limit of detection was improved by about 200-fold compared with the LCQ conditions. With the aid of the GenoMass program, the 4-mer, 5-mer, 6-mer, 7-mer DNA adducts fragments were found and listed, according to the order of abundance. GenoMass then displayed all base sequences corresponding to these ion masses. **Table 4.4** is an example of the six most intense adduct 4-mer ion masses identified and their possible corresponding base sequences. Although isomeric sequences were separated by LC, we have not yet proceeded to analyze these complex DNA adduct data in detail to identify these isomers by internal fragments and/or MS/MS. Nonetheless, these preliminary data demonstrated that the methodology has the potential for identification of AAF bonding sites and to study the relationship of adduct bonding in relation to the location of an adduction site along a DNA chain in a sequence context. We have included representative 4-mer, 5-mer, 6-mer, 7-mer, 8-mer and 9-mer adducts data here.

Table 4.4. The 6 most abundant AAF modification products of the 4-mer fragments [**x**] found on the ds ctDNA. m/z values are those of the AAF adducted 4-mers listed.

m/z	1354.8	4	GCCC	CGCC	CCGC	CCCG								
m/z	1417.9	12	AATG	AAGT	ATAG	ATGA	AGAT	AGTA	TAAG	TAGA	TGAA	GAAT	GATA	GTAA
m/z	1393.8	24	ATGC	ATCG	AGTC	AGCT	ACTG	ACGT	TAGC	TACG	TGAC	TGCA	TCAG	TCGA
			GATC	GACT	GTAC	GTCA	GCAT	GCTA	CATG	CAGT	CTAG	CTGA	CGAT	CGTA
m/z	1418.8	12	AGGC	AGCG	ACGG	GAGC	GACG	GGAC	GGCA	GCAG	GCGA	CAGG	CGAG	CGGA
m/z	1408.8	12	ATTG	ATGT	AGTT	TATG	TAGT	TTAG	TTGA	TGAT	TGTA	GATT	GTAT	GTTA
m/z	1424.8	6	TTGG	TGTG	TGGT	GTTG	GTGT	GGTT							

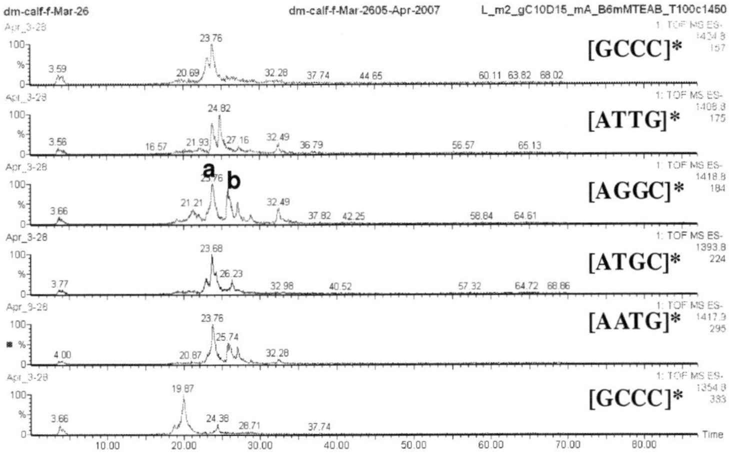

Figure 4.12a. EICs of identified 6 most abundant AAF-modified 4-mers in the digest of calf thymus DNA following exposure to AAAF. Automatically generated and plotted by GenoMass. [**x**] denotes the compositional isomers.

For the most abundant compounds in the chromatograms, the isomeric peaks were differentiated by MS/MS. For example, a and b in the third panel of **Figure 4.12a** which shows isomers of [AGGC]*, i.e, A*GGC or AG*GC could be told apart by their characteristic MS/MS spectra, with the major difference in the [A*G-H]- ion, as shown in **Figure 4.12b**.

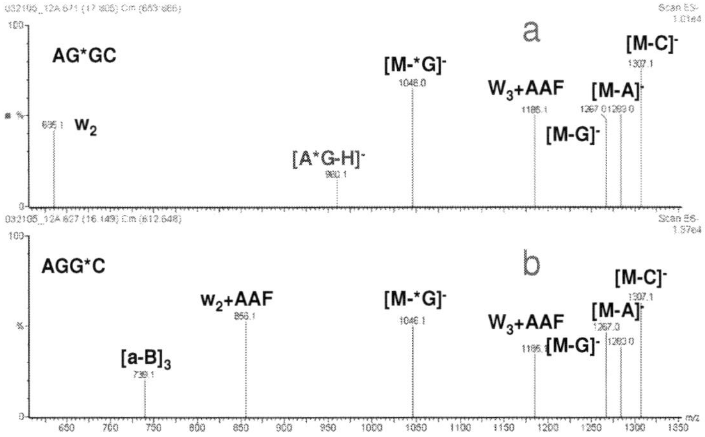

Figure 4.12b. MS/MS spectra of two AAF-modified isomeric 4-mers AG*GC and AGG*C in the digest of calf thymus DNA following exposure to AAAF. Denoted as **a** and **b** in the third panel of **Figure 4.11a**.

In the 5-mer, 6-mer, 7-mer, 8-mer, and 9-mer digestion products studies (**Figure 4.12(c-g)**), we did not conduct MS/MS, as actually the sequence of separated peaks can be identified by the strategy we described before, which is that by accurate MS and characteristic MS spectra the sequence could be deduced and verified.

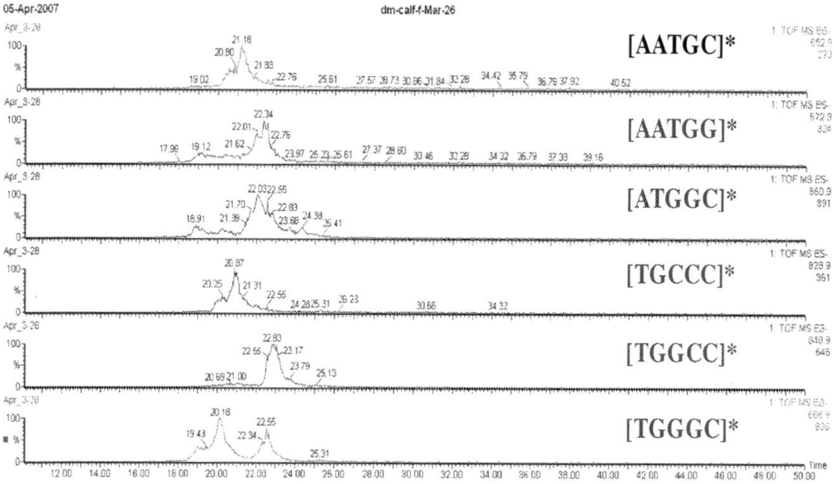

Figure 4.12c. EICs of identified 6 most abundant AAF-modified 5-mers in the digest of ctDNA following exposure to AAAF. Automatically generated and plotted by GenoMass. [**x**] denotes the compositional isomers.

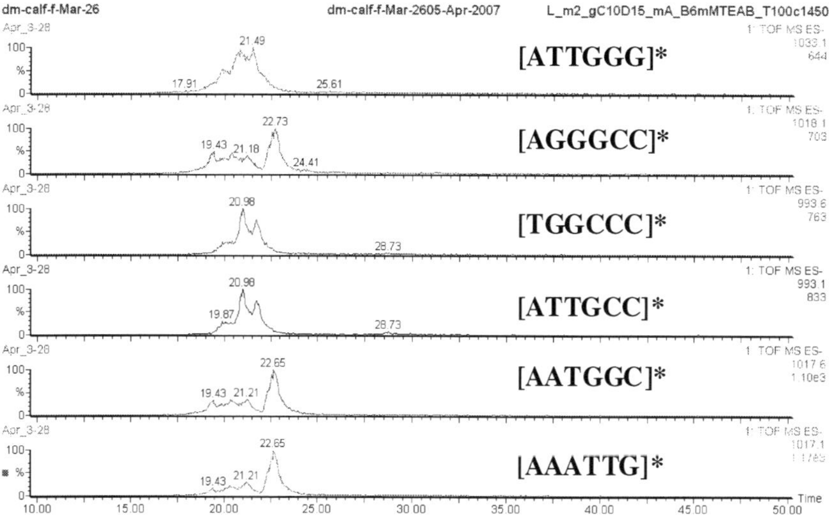

Figure 4.12d. EICs of identified 6 most abundant AAF-modified 6-mers in the digest of ctDNA following exposure to AAAF. Automatically generated and plotted by GenoMass. [**x**] denotes the compositional isomers.

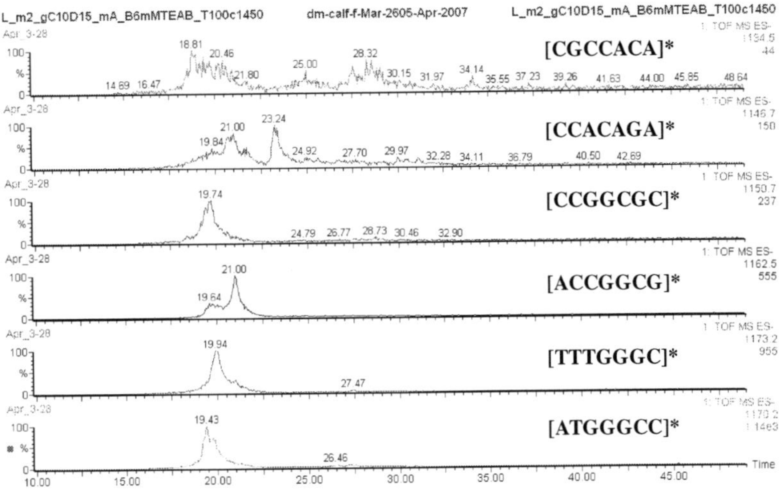

Figure 4.12e. EICs of identified 6 most abundant AAF-modified 7-mers in the digest of ctDNA following exposure to AAAF. Automatically generated and plotted by GenoMass. [**x**] denotes the compositional isomers.

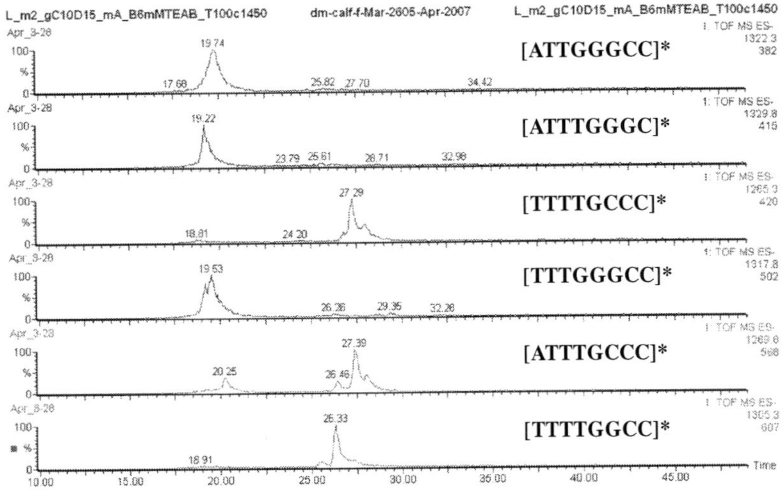

Figure 4.12f. EICs of identified 6 most abundant AAF-modified 8-mers in the digest of ctDNA following exposure to AAAF. Automatically generated and plotted by GenoMass. [**x**] denotes the compositional isomers.

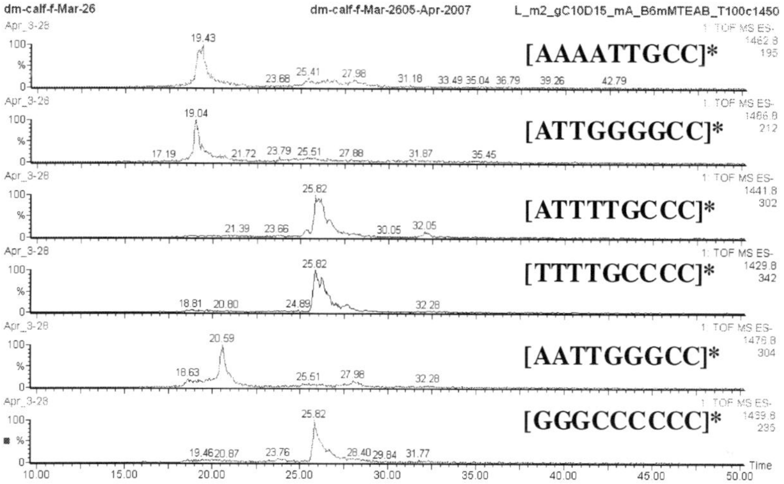

Figure 4.12g. EICs of identified 6 most abundant AAF-modified 9-mers in the digest of ctDNA following exposure to AAAF. Automatically generated and plotted by GenoMass. [**x**] denotes the compositional isomers.

We can draw the conclusion based on the above data that the mutational hotspot may not necessarily correspond to the chemical hotspot. The possible explanation of the sequence specificity: 1) G in different sequence contexts have inherently different activities, or 2) the reactive chemical has specific non-covalent interactions with the DNA double helix that vary with sequence and lead to differences in the subsequent covalent interactions with those sequences.

4.4 Conclusions

Covalent modification of DNA by carcinogens or their metabolites represents an early critical step in multistage chemical carcinogenesis. Cigarette smoking, environmental tobacco smoke, exhaust by-products, food, etc., may all contribute to a person's cumulative risk of cancer. If not repaired before DNA replication, DNA adducts can cause mispairing, resulting in mutation and chromosomal alteration. One of the human cancer theories is based on the idea that a mutation, induced by a covalent modification of DNA, may be an initial stage in chemical carcinogenesis. The exposure to most carcinogens results in damage to the structural integrity of DNA, which most likely results from covalent binding of reactive metabolites of carcinogens to DNA, leading to the formation of DNA adducts. The majority of DNA adducts is eliminated by DNA repair processes; however, some persistent adducts often cause mutation in important growth-controlling genes or loci, resulting in aberrant cellular growth and cancer. Alkylation at C-8 of guanine residues by the heptacarcinogen AAF leads to significant disruption of the double helix around the adduct. The rationale for measuring carcinogen DNA adducts derives from the fact that a number of chemical carcinogens or their metabolites exert their biological effects by binding covalently to cellular DNA, thereby inducing mutations in the critical cellular genes. There is no simple algorithm for translating DNA adduct levels into cancer risk. The relative roles of chemical reactivity in the formation of a lesion and the carcinogenic potency of a particular lesion in the establishment of colonal growth advantage remain enigmatic.

Recent studies strongly suggest that both preferential adduct formation and slow DNA-repair lead to such mutational hotspots. Different sequence contexts induce different DNA distortion. The determination of the adduct sites, especially the isomeric adducts, is a very important premise to learn which site of modification within a target sequence should ascribed to the mutagenic activities. Hence it is important to investigate the sequence's susceptibility to the adduction of AAF. The biological effects of possible mutation induced by the modifications are most likely a function of the base sequence.

The modification, digestion and ion-pairing LC-MS conditions to study AAF-oligonucleotide adducts have been developed. Semi-quantitative analysis of oligonucleotide fragments gave information on possible sequence preference of AAF- adduction. The study on a NarI sequence known to contain a mutational hotspot suggested that the hotspot was not necessarily the most reactive site. The study of ctDNA indicated sequential preference of adduction. The novel computer program GenoMass automated the data analysis that was too complicated to be accomplished by manual analysis currently performed with most mass spectrometry application software such as MassLynx and LCQ Navigator.

Chapter 5

Future Research Perspectives

5.1 Future Prospects of the GenoMass Program

The research presented in this work has introduced a new approach for determination of DNA damage from environmental carcinogens by focusing on the characterization of DNA adducts in their oligonucleotide form as opposed to the thus far more "traditional" methods which have dealt with the analysis of monomeric adducts. It is hoped that placement of carcinogen adduct formation in the context of the base sequence may, ultimately, determine the relationship between adduction and mutation site(s) in specific gene sequences. At this initial stage of this research, two key issues had to be addressed and dealt with in order to bring this goal closer to fruition: first, enzymatic digestion of DNA into fragments of size suitable for MS analysis and, second, a method for processing the vast amount of data produced from the LC-MS analysis of the DNA digest in order to identify carcinogen modified fragments.

As shown in the preceding chapters, the use of a "reverse pseudo-combinatorial" approach allows the reconstruction of chromatographic profiles by simple calculation of the masses of all possible combinations of the four DNA bases and any carcinogen(s) of interest. In Chapter 4, examples are given which demonstrate that, for an oligomer of known sequence, the GenoMass software can place the carcinogen adduction site in the context of base sequence by the determination of the base composition of the enzymatic digestion products. It is envisioned that these fundamental approaches can be expanded in the future in conjunction with the incorporation of tandem mass spectrometry, high resolution mass spectrometry (exact mass measurement) and emerging capillary chromatographic methods. It is expected that such advances may enable reconstruction of the original digested oligomer(s) and ultimately identify adduction sites of carcinogens in specific gene sequences of genomic DNA. Plausible strategies toward these goals are proposed here.

(a) Determination of Adduction Site in Oligomer of Unknown Sequence by Screening Digested Fragmenst via the GenoMass Software.

Bearing in mind that many enzymatic digestion processes utilize restriction enzymes which produce fragments of finite size, we would expect to initially determine the MW, and perhaps base composition of the large fragment by matrix assisted laser desorption ionization (MALDI) and high resolution mass spectrometry (see below for details). This will be followed by digestion of the large oligomer based on the procedures outlined in this book. The idea is that, for an unknown DNA, we can first map all the different size digestion products, i.e., 2-mers, 3-mers, 4-mers, 5-mers, 6-mers, 7-mers, 8-mers. Then it is possible that by combining all these information, looking at the overlap of fragments to fill in the gaps and by mathematical means, the sequence of the unknown DNA can be postulated.

To illustrate this point we will use here as an example the NarI 12-mer 5'-ACCGGCGCCACA-3' by treating it as an unknown sequence. As stated above, initial MALDI analysis will determine the MW and establish that it was a 12-mer of the indicated composition. Using GenoMass, we then could generate a table of possible compositional isomers of that mass to find that it was a 12-mer. Then, after the sample has gone through the enzymatic digestion process as described before, all the fragments ranging from 2-mer to 5-mers were found and listed (**Table 5.1**). Very similar to solving a puzzle, we could come up with the order and arrangements of the bases of the unknown 12-mer, especially with relation to the adduct bonding site, the sequence information could be deduced.

Table 5.1. n-mer fragments identified by GenoMass in an unknown (12-mer) sequence.

Oligomers Identified	**[M – H]⁻**	**Sequences**
2-mer	555.4	GC CG
	539.4	AC CA
	595.4	GG
3-mer	852.6	ACA
	828.6	ACC CCA
	844.6	GCC CCG CGC
	884.6	GGC CGG
4-mer	1141.8	CACA
	1157.8	ACCG
	1173.8	GCGC
5-mer	1431.0	CCACA
	1487.0	ACCGG

Ladders:

4-mer	ACCG GCGC CACA
2-mer	AC CG GC GC (CA)
5-mer	ACCGG ↓ CCACA
3-mer	ACC GGC GCC ACA

The overlap of the n-mers was then used to fill in the missing gaps to eventually postulate the entire sequence to be 5'-ACCG GCGC CACA-3'.

The methodology needs to be further tested with longer oligonucleotides and eventually to see if it is applicable to a "real" DNA adduct sample. The idea is not to sequence the whole DNA, rather would just need to find out the base sequence information surrounding the interested adduct binding site to study the DNA in the sequence context.

This above approach may or may not need the assistance and corroboration from identification of isomers of the digested fragments by LC-MS/MS. However, an additional feature that will augment the information base will be the incorporation of MS/MS data generation into the GenoMass program for isomer differentiation. This can be done by having GenoMass take real-time instrument control to automatically start MS/MS data acquisition with found ions of interest. These extra layers will require substantial additional computational power and data processing capacity but will result in the development of a more seamless program for the sequencing of oligonucleotide adducts. Some of the existing DNA sequencing techniques, mostly developed by J. A. McCloskey and co-workers, will be useful in this regard (Rozenski and McCloskey 1999, 2002; Oberacher et al., 2002).

(b) Determination of Oligonucleotide Adduct Sequence by Accurate Mass Measurements

Accurate mass measurements, aka, high-resolution MS (HRMS) have been a routine method in the determination of the elemental composition of small molecules in analytical laboratories. In fact, the value of accurate mass determination in limiting the possible base compositions of an oligonucleotide cannot be overstated. Since DNA is only made up of 4 mononucleotides A, T, G, C, unlike proteins which may be made up with as many as 20 different amino acids, DNA composition determination should prove to be easier than that for peptides. In addition, the nucleotides do not have any isobaric residues like the amino acids lysine/asparagines and leucine/isoleucine. Thus, accurate mass measurement can play a pivotal role in DNA composition assignments.

Figure 5.1 illustrates the four subunits of DNA with their accurate mass numbers.

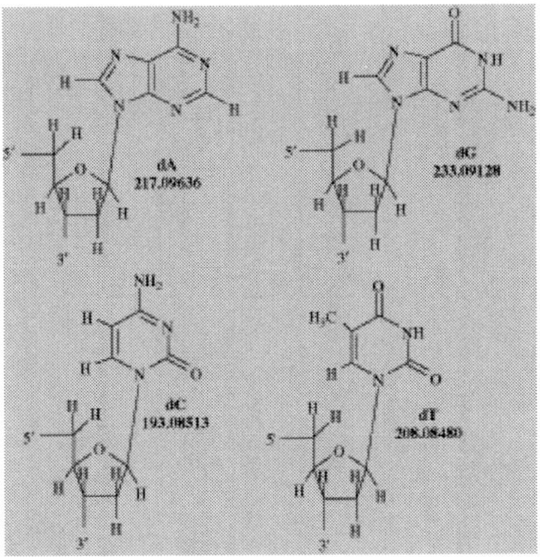

Figure 5.1. The structures and monoisotopic masses of the DNA nucleotide residues (Koomen et al., 2002).

The identification of nucleotide compositions by accurate MS alone becomes increasingly difficult with increasing oligomer length and the number of compositional isomers increases exponentially with n. The number of possible different compositions is:

$$\mathbf{N = 4^n}$$

Where **n** is the number of nucleotides or chain length.
The possible unique masses for oligmers of different size (chain length) are:

$$\mathbf{M = (4+n-1)!/n!/3!}$$

Therefore for 2-mers, there are $N = 4^2 = 16$ different compositions and $M = 10$ unique masses; For 3-mers, there are $N = 4^3 = 64$ different compositions and $M = 20$ unique masses; For 4-mers, there are $N = 4^4 = 256$ different compositions and $M = 35$ unique masses; For 5-mers, there are $N = 4^5 = 1024$ different compositions and $M = 56$ unique masses. The numbers increase exponentially and make it a daunting task to analyze the data. **Figure 5.2** demonstrates the identification of oligonucleotide composition as a function of mass accuracy and the oligomer length.

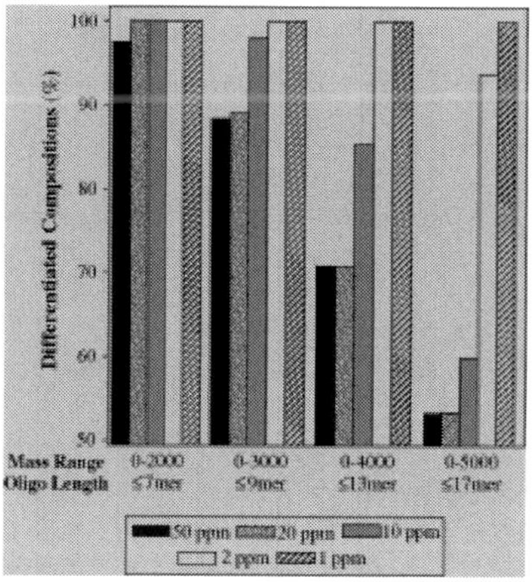

Figure 5.2. Composition identification as a function of mass measurement accuracy for nucleic acids over increasing mass ranges. The corresponding oligonucleotide length is listed below each mass range. The percentage resolved at 50 ppm mass accuracy is at the left and the percentage resolved at 1 ppm mass accuracy is on the right of each grouping (Koomen et al., 2002).

In the model systems studied in this work, mass measurements were conducted with low or moderate resolution instruments (Ion Trap LCQ or Reflector Time of Flight) where mass accuracy was in the range of 10 to 20 ppm. As a result, assignment of base composition for segments of 4-mers to 7-mer in length, as discussed in this thesis, presented no major challenge. The same fundamental concepts can now be pursued by simple expansion of computer power that will allow reliable access of oligonucleotide adducts of larger length to eventually identify the adduction sites of specific gene sequences in genomic DNA.

As suggested by one of my colleagues (Dr. James Glick) in an independent discussion, an important improvement to the software should be the development of an interface that would use the identified oligonucleotide output from GenoMass to interrogate the Human Genome database. The National Center for Biotechnology Information (NCBI) at the National Institutes of Health (NIH) provides an interface for the BLAST (Basic Local Alignment Search Tool) searching of the human genome using either a nucleotide basepair length of 7-20 or greater than 20 basepairs. Presently, GenoMass can extract data for oligonucleotide fragments up to 10 basepairs in length but this length is too short to provide meaningful BLAST results for specific genes when a manual BLAST search is performed. Further modification of the software and the use of MS systems capable of handling higher molecular weight oligomers as discussed in this section could result in the development of an automated searching interface to perform Human Genome BLAST searches. This feature would be analogous to proteomics based software such as Sequest™ which uses the MS/MS data from peptides to perform the automated sequencing of the peptide and then search the NCBI database to identify the related protein. With a fully developed GenoMass interface, the identification of an adducted nucleotide sequence(s) could be automatically used to search the Human Genome database to identify specific genes susceptible to adduction or to develop a genomic wide "fingerprint" of DNA adduction.

REFERENCES

Abdel-Baky, S.; Giese, R. W. Gas Chromatography/Electron Capture Negative-Ion Mass Spectrometry at the Zeptomole Level. *Anal. Chem.* **1991,** *63*, 2986-2989.

Abian, J.; Oosterkamp, A. J.; Gelpi, E. Comparison of Conventional, Narrow-Bore and Capillary Liquid Chromatography/Mass Spectrometry for Electrospray Ionization Mass Spectrometry: Practical Consideration. *J. Mass Spectrom.* **1999,** *34*, 244-254.

Allam, K.; Abdel-Baky, S.; Giese, R. W. Potassium Superoxide Chemical Transformation/Mass Spectrometry Detection of Covalent Damage to the DNA of Cultured Human Lymphocytes Exposed to Benzo[a]pyrene. *Anal. Chem.* **1993,** *65*, 1723-1727.

Andrews, C. L.; Vouros, P.; Harsch, A. Analysis of DNA Adducts Using High-Performance Separation Techniques Coupled to Electrospray Ionization Mass Spectrometry. *J. Chromatogr., A* **1999,** *856*, 515-526.

Andrews, C. L.; Harsch, A.; Vouros, P. Analysis of the in Vitro Digestion of Modified DNA to Oligonucleotides by LC-MS and LC-MS/MS. *Int. J. Mass Spectrom.* **2004,** *231*, 169-177.

Apffel, A.; Chakel, J. A.; Fischer, S.; Lichtenwalter, K.; Hancock, W. S. Analysis of Oligonucleotides by HPLC-Electrospray Ionizaiton Mass Spectrometry. *Anal. Chem.* **1997a,** *69*, 1320-1325.

Apffel, A.; Chakel, J. A.; Fischer, S.; Lichtenwalter, K.; Hancock, W. S. New Procedure for the Use of High Performance Liquid Chromatography-Electrospray Ionizaiton Mass Spectrometry for the Analysis of Nucelotides and Oligonucleotides. *J. Chromatogr., A* **1997b,** *777*, 3-21.

Apruzzese, W. A.; Vouros, P. Analysis of DNA Adducts by Capillary Methods Coupled to Mass Spectrometry: A Perspective. *J. Chromatogr., A* **1998,** *794*, 97-108.

Bailey, E.; Farmer, P. B.; Shuker, D. E. Estimation of Exposure to Alkylating Carcinogens by the GC-MS Determination of Adducts to Hemoglobin and Nucleic Acid Bases in Urine. *Arch. Toxicol.* **1987**, *60*, 187-191.

Bailey, E.; Brooks, A. G.; Bird, I.; Farmer, P. B.; Street, B. Monitoring Exposure to 4,4'-Methylenedianiline by the Gas Chromatography-Mass Spectrometry Determination of Adducts to Hemoglobin. *Anal. Biochem.* **1990**, *190*, 175-181.

Barry, J. P.; Vouros, P.; Van Schepdael, A.; Lay, S. J. Mass and Sequence Verification of Modified Oligonucleotides Using Electrospray Mass Spectrometry. *J. Mass Spectrom.* **1995**, *30*, 993-1006.

Barry, J. P.; Muth, J.; Law, S. J.; Karger, B. L.; Vouros, P. Analysis of Modified Oligonucleotides by Capillary Electrophoresis in a Polyvinylpyrrolidone Matrix Coupled with Electrospray Mass Spectrometry. *J. Chromatogr., A.* **1996a,** *732*, 159-166.

Barry, J. P.; Norwood, C.; Vouros, P. Detection and Identification of Benzo[a]pyrene Diol Epoxide Adducts to DNA Utilizing Capillary Electrophoresis-Electrospray Mass Spectrometry. *Anal. Chem.* **1996b,** *68*, 1432-1438.

Basu, A. K.; Essigmann, J. M. Site-Specifically Modified Oligodeoxynucleotides as Probes for the Structural and Biological Effects of DNA-Damaging Agents. *Chem. Res. Toxicol.* **1988**, *1*, 1-18.

Beland, F. A.; Churchwell, M. I.; Von Tungeln, L. S.; Chen, S.; Fu, P. P.; Culp, S. J.; Schoket, B.; Gyorffy, E.; Minarovits, J.; Poirier, M. C.; Bowman, E. D.; Weston, A.; Doerge, D. R. High-Performance Liquid Chromatography Electrospray Ionization

Tandem Mass Spectrometry for the Detection and Quantization of Benzo[a]pyrene-DNA Adducts. *Chem. Res. Toxicol.* **2005,** *18*, 1306-1315.

Belguise-Valladier, P.; Fuchs, R. P. P. Strong Sequence-Dependent Polymorphism in Adduct-Induced DNA Structure Analysis of Single N-2-Acetylaminofluorene Residues Bound within the NarI Mutation Hot Spot. *Biochemistry* **1991,** *30*, 10091-10100.

Bintz, R.; Fuchs, R. P. P. Induction of -2 Frameshift Mutations within Alternating GC Sequences by Carcinogens That Bind to the C8 Position of Guanine Residues: Development of a Specific Mutation Assay. *Mol. Gen. Genet.* **1990**, *221*, 331-338.

Bleicher, K.; Bayer, E. Various Factors Influencing the Signal Intensity of Oligonucleotides in Electrospray Mass Spectrometry. *Biol. Mass Spectrom.* **1994a,** *23*, 320-322.

Bleicher, K.; Bayer, E. Analysis of Oligonucleotides Using Coupled High Performance Liquid Chromatography-Electrospray Mass Spectrometry. *Chromatographia* **1994b,** *39*, 405-408.

Bohr, V. A. Gene Specific DNA Repair. *Carcinogenesis* **1991,** *12*, 1983-1992.

Borges, C.; Lemiere, F.; Embrechts, J.; Van Dongen, W.; Esmans, E. L. Characterisation of Estrone-Nucleic Acid Adducts Formed by Reaction of 3,4-Estrone-O-Quinone with 2'-Deoxynucleosides/Deoxynucleotides Using Capillary Liquid Chromatography/Electrospray Ionization Mass Spectrometry. *Rapid Commun. Mass Spectrom.* **2004,** *18*, 2191-2200.

Bothner, B.; Chatman, K.; Sarkisian, M.; Siuzdak, G. Liquid Chromatography Mass Spectrometry of Antisense Oligonucleotides *Bioorg. Med. Chem. Lett.* **1995,** *5*, 2863-2868.

Bowers, M. T. Ion Mobility: The Method and Some Applications to Oligonucleotides and Peptides/Proteins. *Abstracts of Papers,* 17th International Mass Spectrometry Conference, Prague, 2006.

Brown, K.; Harvey, C. A.; Turteltaub, K. W.; Shields, S. J. Structural Characterization of Carcinogen-Modified Oligodeoxynucleotide Adducts Using Matrix-Assisted Laser Desorption/ionization Mass Spectrometry *J. Mass Spectrom.* **2003,** *38*, 68-79.

Bryant, M. S.; Skipper, P. L.; Tannenbaum, S. R.; Maclure, M. Hemoglobin Adducts of 4-Aminobiphenyl in Smokers and Nonsmokers. *Cancer Res.* **1987**, *47*, 602-608.

Bryant, M. S.; Vineis, P.; Skipper, P. L.; Tannenbaum, S. R. Hemoglobin Adducts of Aromatic Amines: Associations with Smoking Status and Type of Tobacco. *Proc. Natl. Acad. Sci. U.S.A.* **1988**, *85*, 9788-9791.

Burnouf, D.; Koehl, P.; Fuchs, R. P. P. Single Adduct Mutagenesis: Strong Effect of the Position of a Single Acetylaminofluorene Adduct within a Mutation Hot Spot. *Proc. Natl. Acad. Sci. U.S.A.* **1989**, *86*, 4147-4151.

Carmella, S. G.; Kagan, S. S.; Kagan, M.; Foiles, P. G.; Palladino, G.; Quart, A, M.; Quart, E.; Hecht, S. S. Mass Spectrometric Analysis of Tobacco-Specific Nitrosamine Hemoglobin Sdducts in Snuff Dippers, Smokers, and Nonsmokers. *Cancer Res.* **1990**, *50*, 5438-5445.

Chaudhary, A. K.; Nokubo, M.; Reddy, G. R.; Blair, I. A.; Marnett, L. J. Detection of Endogenous Malondialdehyde-Deoxyguanosine Adducts in Human Liver. *Science* **1994,** *265*, 1580-1582.

Chaudhary, A. K.; Nokubo, M.; Oglesby, T. D.; Marnett, L. J.; Blair, I. A. Characterization of Endogenous DNA Adducts by Liquid Chromatography Tandem Mass Spectrometry *J. Mass Spectrom.* **1995,** *30*, 1157-1166.

Chen, H.; Zhang, W.; Song, R.; Ma, H.; Dong, Y.; Sheng, G.; Zhou, Z.; Fu, J. Analysis of DNA Methylation by Tandem Ion-Pair Reverse-Phase High-Performance Liquid Chromatography/Electrospray Ionization Mass Spectrometry. *Rapid Commun. Mass Spectrom.* **2004,** *18*, 2773-2778.

Chen, H.-J. C. Formation, Analysis and Repair of Exocyclic Etheno DNA Adducts. *Chin. Pharm. J. (Taipei, Taiwan)* **2004,** *56*, 1-16.

Chen, H.-J. C.; Chang, C.-M. Quantification of Urinary Excretion of 1,N6-Ethenoadenine, a Potential Biomarker of Lipid Peroxidation, in Humans by Stable Isotope Dilution Liquid Chromatography-Electrospray Ionization-Tandem Mass Spectrometry: Comparison with Gas Chromatography-Mass Spectrometry. *Chem. Res. Toxicol.* **2004,** *17*, 963-971.

Chen, H.-J. C.; Chiu, W.-L. Association Between Cigarette Smoking and Urinary Excretion of 1,N2-Ethenoguanine Measured by Isotope Dilution Liquid Chromatography-Electrospray Ionization/Tandem Mass Spectrometry. *Chem. Res. Toxicol.* **2005,** *18*, 1593-1599.

Chen, H.-J. C.; Hong, C.-L.; Wu, C.-F.; Chiu, W.-L. Effect of Cigarette Smoking on Urinary 3, N4-Ethenocytosine Levels Measured by Gas Chromatography/Mass Spectrometry. *Toxicol. Sci.* **2003,** *76*, 321-327.

Chen, H.-J. C.; Zhang, W.; Song, R.; Ma, H.; Dong, G. S.; Zhou, Z.; Fu, J. Analysis of DNA Methylation by Tandem Ion-Pair Reversed-Phase High-Performance Liquid Chromatography/Electrospray Ionization Mass Spectrometry. *Rapid Commun. Mass Spectrom.* **2004a,** *19*, 1120-1124.

Chen, H.-J. C.; Wu, C.-F.; Hong, C.-L.; Chang, C.-M. Urinary Excretion of 3,N4-Etheno-2'-deoxycytidine in Humans as a Biomarker of Oxidative Stress: Association with Cigarette Smoking. *Chem. Res. Toxicol.* **2004b,** *17*, 896-903.

Chen, W.-H.; Qin, Y.; Cai, Z.; Chan, C.-L.; Jiang, Z.-H. Spectrometric Studies of Cytotoxic Protoberberine Alkaloids Binding to Double-Stranded DNA. *Bioorg. Med. Chem.* **2005,** *13*, 1859-1866.

Chen, Y.; Liu, X.; Pisha, E.; Constantinou, A. I.; Hua, Y.; Shen, L.; Van Breemen, R. B.; Elguindi, E. C.; Blond, S. Y.; Zhang, F.; Bolton, J. L. A Metabolite of Equine Estrogens, 4-Hydroxyequilenin, Induces DNA Damage and Apoptosis in Breast Cancer Cell Lines. *Chem. Res. Toxicol.* **2000,** *13*, 342-350.

Chien, R.-L.; Burgi, D. S. Sample Stacking of an Extremely Large Injection Volume in High-Performance Capillary Electrophoresis. *Anal. Chem.* **1992,** *64*, 1046-1050.

Chou, M. W.; Jian, Y.; Williams, L. D.; Xia, Q.; Churchwell, M.; Doerge, D. R.; Fu, P. P. Identification of DNA Adducts Derived from Riddelliine, a Carcinogenic Pyrrolizidine Alkaloid. *Chem. Res. Toxicol.* **2003,** *16*, 1130-1137.

Churchwell, M. I.; Beland, F. A.; Doerge, D. R. Quantification of Multiple DNA Adducts Formed through Oxidative Stress Using Liquid Chromatography and Electrospray Tandem Mass Spectrometry. *Chem. Res. Toxicol.* **2002,** *15*, 1295-1301.

Churchwell, M. I.; Beland, F. A.; Doerge, D. R. Quantification of O6-Methyl and O6-Ethyl Deoxyguanosine Adducts in C57BL/6N/Tk +/- Mice Using LC/MS/MS. *J. Chromatogr., B: Anal. Technol. Biomed. Life Sci.* **2006a,** *844*, 60-66.

Churchwell, M. I.; Yan, J.; Xia, Q.; Fu, P. P.; Beland, F. A.; Doerge, D. R. LC/MS/MS Analysis of Benzo[a]pyrene DNA Adducts Derived from Diol Epoxide and Cation Radical Pathways. *Abstracts of Papers,* 232nd National Meeting of the American

Chemical Society, San Francisco, CA, Sept 10-14, 2006; American Chemical Society: Washington, DC, 2006b; TOXI-052.

Colgrave, M. L.; Iannitti-Tito, P.; Wickham, G.; Sheil, M. M. Rapid Determination of Sequence Selectivity and Stability of Alkylated Oligonucleotide Adducts by Electrospray Tandem Mass Spectrometry. *Aust. J. Chem.* **2003,** *56*, 401-413.

Conesa, A.; Gotz, S.; Garcia-Gomez, J. M.; Terol, J.; Talon, M.; Robles, M. Blast2GO: A Universal Tool for Annotation, Visualization and Analysis in Functional Genomics Research. *Bioinformatics* **2005,** *21*, 3674-3676.

Day, B. W.; Naylor, S.; Gan, L. S.; Sahali, Y.; Nguyen, T. T.; Skipper, P. L.; Wishnok, J. S.; Tannenbaum, S. R. Molecular Dosimetry of Polycyclic Aromatic Hydrocarbon Epoxides and Diol Epoxides via Hemoglobin Adducts. *Cancer Res.* **1990**, *50*, 4611-4618.

De Kok, T. M. C. M.; Moonen, H. J. J.; Van Delft, J.; Van Schooten, F. J. Methodologies for Bulky DNA Adduct Analysis and Biomonitoring of Environmental and Occupational Exposures. *J. Chromatogr., B: Anal. Technol. Biomed. Life Sci.* **2002,** *778*, 345-355.

Debrauwer, L.; Rathahao, E.; Jouanin, I.; Paris, A.; Clodic, G.; Molines, H.; Convert, O.; Fournier, F.; Tabet, J. C. Investigation of the Regio- and Stereo-Selectivity of Deoxyguanosine Linkage to Deuterated 2-Hydroxyestradiol by Using Liquid Chromatography/ESI-Ion Trap Mass Spectrometry. *J. Am. Soc. Mass Spectrom.* **2003,** *14*, 364-372.

Deforce, D. L.; Van den Eeckhout, E. G. Analysis of DNA Damage Using Capillary Zone Electrophoresis and Electrospray Mass Spectrometry. *Methods in Mol. Biol. (Totowa, NJ, U.S.)* **2001,** *162*, 429-441.

Dickman, M. J.; Matin, M. M.; Hornby, D. P. High-Throughput Analysis of Nucleic Acid Modification Reactions Using Ion-Pair Reverse-Phase High-Performance Liquid Chromatography. *Anal. Biochem.* **2002,** *301*, 290-297.

Ding, J.; Vouros, P. Advances in CE/MS. *Anal. Chem.* **1999,** *71*, 378A-385A.

Doerge, D. R.; Yi, P.; Churchwell, M. I.; Preece, S. W.; Langridge, J.; Fu, P. P. Mass Spectrometric Analysis of 2'-Deoxyribonucleoside and 2'-Deoxyribonucleotide Adducts with Aldehydes derived from Lipid Peroxidation. *Rapid Commun. Mass Spectrom.* **1998,** *12*, 1665-1672.

Doerge, D. R.; Churchwell, M. I.; Fang, J.-L.; Beland, F. A. Quantification of Etheno-DNA Adducts Using Liquid Chromatography, On-Line Sample Processing, and Electrospray Tandem Mass Spectrometry. *Chem. Res. Toxicol.* **2000,** *13*, 1259-1264.

Doerge, D. R.; Churchwell, M. I.; Beland, F. A. Analysis of DNA Adducts from Chemical Carcinogens and Lipid Peroxidation Using Liquid Chromatography and Electrospray Mass Spectrometry. *J. Environ. Sci. Health, Part C, Environ. Carcinog. Ecotoxicol. Rev.* **2002,** *20*, 1-20.

Edler, M.; Jakubowski, N.; Linscheid, M. Styrene Oxide DNA Adducts: Quantitative Determination Using 31P Monitoring. *Anal. Bioanal. Chem.* **2005,** *381*, 205-211.

Embrechts, J.; Lemiere, F.; Van Dongen, W.; Esmans, E. L. Equilenin-2'-Deoxynucleoside Adducts: Analysis with Nano-Liquid Chromatography Coupled to Nano-Electrospray Tandem Mass Spectrometry. *J. Mass Spectrom.* **2001,** *36*, 317-328.

Embrechts, J.; Lemiere, F.; Van Dongen, W.; Esmans, E. L.; Buytaert, P.; Van Marck, E.; Kockx, M.; Makar, A. Detection of Estrogen DNA-Adducts in Human Breast Tumor Tissue and Healthy Tissue by Combined Nano LC-Nano ES Tandem Mass Spectrometry. *J. Am. Soc. Mass Spectrom.* **2003,** *14*, 482-491.

Erickson, B. E. Bisphenol A Under Scrutiny. *Chem. Eng. News* **2008**, *86* (22), 36-39.

Esaka, Y.; Inagaki, S.; Goto, M. Separation Procedures Capable of Revealing DNA Adducts. *J. Chromatogr., B: Anal. Technol.Biomed. Life Sci.* **2003,** *797*, 321-329.

Esmans, E. L.; Broes, D.; Hoes, I.; Lemiere, F.; Vanhoutte, K. Liquid Chromatography-Mass Spectrometry in Nucleoside, Nucleotide and Modified Nucleotide Characterization. *J. Chromatogr., A* **1998,** *794*, 109-127.

Fan, J. B.; Bibikova, M. Specific DNA Methylation Patterns for Diagnosing Lung Cancer. Patent 2004-973783, 2005164246, 20041025, 2005.

Farmer, P. B.; Bailey, E. Protein-Carcinogen Adducts in Human Dosimetry. *Arch. Toxicol. Suppl.* **1989**, *13*, 83-90.

Farmer, P. B.; Bailey, E.; Gorf, S. M.; Törnqvist, M.; Osterman-Golkar, S.; Kautiainen, A.; Lewis-Enright, D. P. Monitoring Human Exposure to Ethylene Oxide by the Determination of Haemoglobin Adducts Using Gas Chromatography-Mass Spectrometry. *Carcinogenesis* **1986**, *7*, 637-640.

Farmer, P. B.; Lamb, J.; Lawley, P. D. Novel Uses of Mass Spectrometry in Studies of Adducts of Alkylating Agents with Nucleic Acids and Proteins. *IARC Sci. Publ.* **1988**, *89*, 347-355.

Farmer, P. B.; Brown, K.; Tompkins, E.; Emms, V. L.; Jones, D. J. L.; Singh, R.; Phillips, D. H. DNA Adducts: Mass Spectrometry Methods and Future Prospects. *Toxicol. Appl. Pharmacol.* **2005,** *207*, (Suppl. 2), S293-S301.

Fuchs, R. P. P. DNA Binding Spectrum of the Carcinogen N-Acetoxy-N-2-Acetylacminofluorene Significantly Differs from the Mutation Spectrum. *J. Mol. Biol.* **1984,** *177*, 173-180.

Gaddis, S. S.; Wu, Q.; Thames, H. D.; DiGiovanni, J.; Walborg, E. F.; MacLeod, M. C.; Vasquez, K. M. A Web-Based Search Engine for Triplex-Forming Oligonucleotide Target Sequences. *Oligonucleotides* **2006,** *16*, 196-201.

Gangl, E.; Turesky, R.; Vouros, P. Detection of in Vivo Formed DNA Adducts at the Part Per Billion Level by Capillary Liquid Chromatography/Microelectrospray Mass Spectrometry. *Anal. Chem.* **2001**, *73*, 2397-2404.

Gao, L.; Zhang, L.; Cho, B. P.; Chiarelli, M. P. Sequence Verification of Oligonucleotides Containing Multiple Arylamine Modifications by Enzymatic Digestion and Liquid Chromatography Mass Spectrometry (LC/MS). *J. Am. Soc. Mass Spectrom.* **2008,** *19*, 1147-1155.

Garner, R. C. The Role of DNA Adducts in Chemical Carcinogenesis. *Mutat. Res.* **1998,** *402*, 67-75.

Gennaro, L. A.; Vadhanam, M.; Gupta, R. C.; Vouros, P. Selective Digestion and Novel Cleanup Techniques for Detection of Benzo[a]pyrene Diol Epoxide-DNA Adducts by Capillary Electrophoresis/Mass Spectrometry. *Rapid Commun. Mass Spectrom.* **2004,** *18*, 1541-1547.

Giese, R. W. Electron-Capture Mass Spectrometry: Recent Advances. *J. Chromatogr., A* **2000,** *892*, 329-346.

Giese, R. W.; Saha, M.; Abdel-Baky, S.; Allam, K. Measuring DNA Adducts by Gas Chromatography-Electron Capture-Mass Spectrometry: Trace Organic Analysis. *Methods Enzymol.* **1996,** *271*, 504-522.

Glover, R. P.; Roberts, G. C. K.; Sweetman, G. M. A.; Peter, F. B. Sequencing of Oligonucleotides Using High Performance Liquid Chromatography and Electrospray Mass Spectrometry. *Rapid Commun. Mass Spectrom.* **1995,** *9*, 897-901.

Glover, R. P.; Lamb, J. H.; Farmer P. B. Tandem Mass Spectrometry Studies of a Carcinogen Modified Oligonucleotide. *Rapid Commun. Mass Spectrom.* **1998,** *12*, 368-372.

Gonzalez-Reche, L. M.; Koch, H. M.; Weiss, T.; Muller, J.; Drexler, H.; Angerer, J. Analysis of Ethenoguanine Adducts in Human Urine Using High Performance Liquid Chromatography-Tandem Mass Spectrometry. *Toxicol. Lett.* **2002,** *134*, 71-77.

Guo, B. Mass Spectrometry in DNA Analysis. *Anal. Chem.*, **1999**, *71*, 333R-337R.

Guza, R.; Rajesh, M.; Fang, Q.; Pegg, A. E.; Tretyakova, N. Kinetics of O6-Methyl-2'-Deoxyguanosine Repair by O6-Alkylguanine DNA Alkyltransferase within K-Ras Gene-Derived DNA Sequences. *Chem. Res. Toxicol.* **2006,** *19*, 531-538.

Haglund, J.; Henderson, A. P.; Golding, B. T.; Tornqvist, M. Evidence for Phosphate Adducts in DNA from Mice Treated with 4-(N-Methyl-N-Nitrosamino)-1-(3-Pyridyl)-1-Butanone (NNK). *Chem. Res. Toxicol.* **2002,** *15*, 773-779.

Hail, M. E.; Elliott, B.; Anderson, J. High-Throughput Analysis of Oligonucleotides Using Automated Electrospray Ionization Mass Spectrometry. *Am. Biotechnol. Lab.* **2004**, *22*, 12-14.

Ham, A.-J. L.; Engelward, B. P.; Koc, H.; Sangaiah, R.; Meira, L. B.; Samson, L. D.; Swenberg, J. A. New Immunoaffinity-LC-MS/MS Methodology Reveals that Aag Null Mice Are Deficient in Their Ability to Clear 1,N6-Etheno-Deoxyadenosine DNA Lesions from Lung and Liver in Vivo. *DNA Repair* **2004,** *3*, 257-265.

Harsch, A.; Vouros, P. Interfacing of CE in a PVP Matrix to Ion Trap Mass Spectrometry: Analysis of Isomeric and Structurally Related (N-Acetylamino) Fluorene-Modified Oligonucleotides. *Anal. Chem.* **1998,** *70*, 3021-3027.

Harsch, A.; Marzilli, L. A.; Vouros, P. Separation and Online Identification of Modified Oligonucleotides Using HPLC and CE-ITMS. *Abstracts of Papers,* 215th National Meeting of the American Chemical Society, Dallas, TX, March 29-April 2, 1998; American Chemical Society: Washington, DC, 1998; ANYL-004.

Harsch, A.; Sayer, J. M.; Jerina, D. M.; Vouros, P. HPLC-MS/MS Identification of Positionally Isomeric Benzo[c]phenanthrene Diol Epoxide Adducts in Duplex DNA. *Chem. Res. Toxicol.* **2000,** *13*, 1342-1348.

Hartley, J. A.; Bingham, J. P.; Souhami, R. L. DNA Sequence Selectivity of Guanine-N7 Alkylation by Nitrogen Mustards Is Preserved in Intact Cells. *Nucleic Acids Res.* **1992,** *20*, 3175-3178.

Hecht, S. S.; Kagan, S. S.; Kagan, M.; Carmella, S. G. Quantification of 4-Hydroxy-1-(3-Pyridyl)-1-Butanone Released from Human Haemoglobin as a Dosimeter for Exposure to Tobacco-Specific Nitrosamines. *IARC Sci. Publ.* **1991**, *105*, 113-118.

Hecht, S. S.; Villalta, P. W.; Sturla, S. J.; Cheng, G.; Yu, N.; Upadhyaya, P.; Wang, M. Identification of O2-Substituted Pyrimidine Adducts Formed in Reactions of 4-(Acetoxymethylnitrosamino)-1-(3-Pyridyl)-1-Butanone and 4-(Acetoxymethylnitros-Amino)-1-(3-Pyridyl)-1-Butanol with DNA. *Chem. Res. Toxicol.* **2004,** *17*, 588-597.

Hemminki, K. DNA Adducts, Mutations and Cancer. *Carcinogenesis* **1993,** *14*, 2007-2012.

Hemminki, K.; Koskinen, M.; Rajaniemi, H.; Zhao, C. DNA Adducts, Mutations, and Cancer. *Regul. Toxicol. Pharmacol.* **2000,** *32*, 264-275.

Hillem, B. More Concerns Over Bisphenol A. *Chem. Eng. News* **2007a**, *85* (32), 8.

Hillem, B. Bisphenol A Vexations. *Chem. Eng. News* **2007b**, *85* (36), 31-33.

Hillestrom, P. R.; Hoberg, A.-M.; Weimann, A.; Poulsen, H. E. Quantification of 1,N6-Etheno-2'-Deoxyadenosine in Human Urine by Column-Switching LC/APCI-MS/MS. *Free Radical Biol. Med.* **2004,** *36*, 1383-1392.

Hillestrom P. R.; Nyssonen, K.; Tuomainen, T.-P.; Pukkala, E.; Salonen Jukka, T.; Poulsen Henrik, E. Urinary Excretion of Epsilond A Is Not Predictive of Cancer Development: A Prospective Nested Case-Control Study. *Free Radical Res.* **2005,** *39*, 51-53.

Hillestrom, P. R.; Covas, M.-I.; Poulsen, H. E. Effect of Dietary Virgin Olive Oil on Urinary Excretion of Etheno-DNA Adducts. *Free Radical Biol. Med.* **2006a,** *41*, 1133-1138.

Hillestrom, P. R.; Weimann, A.; Jensen, C. B.; Storgaard, H.; Vaag, A. A.; Poulsen, H. E. Consequences of Low Birth Weight On Urinary Excretion of DNA Markers of Oxidative Stress in Young Men. *Scand. J. Clin. Lab. Invest.* **2006b,** *66*, 363-370.

Hillestrom, P. R.; Weimann, A.; Poulsen, H. E. Quantification of Urinary Etheno-DNA Adducts by Column-Switching LC/APCI-MS/MS. *J. Am. Soc. Mass Spectrom.* **2006c,** *17*, 605-610.

Hoes, I.; Dongen, W.; Vanhoutte, K.; Esmans, E.; Bockstele, D.; Berneman, Z.; Deforce, D.; Van den Eeckhout, E. J. Analysis of Melphlan Adducts of 2-Deoxynucleotides in Calf Thymus DNA Hydrolysates by Capillary High-Performance Liquid Chromatography-Electrospary Tandem Mass Spectrometry. *J. Chromatogr., B: Anal. Technol. Biomed. Life Sci.* **1999**, *736*, 43-59.

Hoes, I.; Van Dongen, W.; Lemiere, F.; Esmans, E. L.; Van Bockstaele, D.; Berneman, Z. N. Comparison Between Capillary and Nano Liquid Chromatography-Electrospray Mass Spectrometry for the Analysis of Minor DNA-Melphalan Adducts. *J. Chromatogr., B: Biomed. Sci. Appl.* **2000,** *748*, 197-212.

Hofer, T.; Badouard, C.; Bajak, E.; Ravanat, J.-L.; Mattsson, A.; Cotgreave, I. A. Hydrogen Peroxide Causes Greater Oxidation in Cellular RNA Than in DNA. *Biol. Chem.* **2005,** *386*, 333-337.

Hoffmann, G. R.; Fuchs, R. P. P. Mechanisms of Frameshift Mutations: Insight from Aromatic Amines. *Chem. Res. Toxicol.* **1997,** *10*, 347-359.

Huber, C. G.; Krajete, A. Analysis of Nucleic Acids by Capillary Ion-Pair Reverse-Phase HPLC Coupled to Negative Ion-Electrospray Ionization Mass Spectrometry. *Anal. Chem.* **1999,** *71*, 3730-3739.

Huber, C. G.; Oberacher, H. Analysis of Nucleic Acids by On-line Liquid Chromatography-Mass Spectrometry. *Mass Spectrom. Rev.* **2002,** *20*, 310-343.

Hunt, D. F.; Crow, F. W. Electron Capture Negative Ion Chemical Ionization Mass Spectrometry. *Anal. Chem.* **1978,** *50*, 1781-1784.

Hunt, D. F.; Stafford, G. C.; Crow, F. W.; Russell, J. W. Pulsed Positive Negative Ion Chemical Ionization Mass Spectrometry. *Anal. Chem.* **1976,** *48*, 2098-2104.

Iannitti, P.; Sheil, M. M.; Wickham, G. High Sensitivity and Fragmentation Specificity in the Analysis of Drug-DNA Adducts by Electrospray Tandem Mass Spectrometry. *J. Am. Chem. Soc.* **1997,** *119*, 1490-1491.

Janning, P.; Schrader, W.; Linscheid, M. A New Mass Spectrometric Approach to Detect Modifications in DNA. *Rapid Commun. Mass Spectrom.* **1994,** *8*, 1035-1040.

Jayaraj, S.; Reid, R.; Santi, D. V. GeMS: An Advanced Software Package for Designing Synthetic Genes. *Nucleic Acids Res.* **2005**, *33*, 3011-3016.

Jian, W.; Lee, S. H.; Blair, I. A. Lipoxygenase Mediated Endogenous DNA Damage. *Abstracts of Papers,* 228th National Meeting of the American Chemical Society, Philadelphia, PA, Aug 22-26, 2004; American Chemical Society: Washington DC, 2004; TOXI-010.

Kaddiska, M. B.; Gladen, B. C.; Baird, D. D.; Germolec, D.; Graham, L. B.; Parker, C. E.; Nyska, A.; Wachsman, J. T.; Ames, B. N.; Basu, S.; Brot, N.; FitzGerald, G. A.; Floyd, R. A.; George, M.; Heinecke, J. W.; Hatch, G. E.; Hensley, K.; Lawson, J. A.; Marnett, L. J.; Morrow, J. D.; Murray, D. M.; Plastaras, J.; Roberts, L. J.; Rokach, J.; Shigenaga, M. K.; Sohal, R. S.; Sun, J.; Tice, R. R.; Van Thiel, D. H.; Wellner, D.; Walter, P. B.; Tomer, K. B.; Mason, R. P.; Barrett, J. C. Biomarkers of Oxidative Stress Study II: Are Oxidation Products of Lipids, Proteins, and DNA Markers of CCl4 Poisoning? *Free Radical Biol. Med.* **2005,** *38*, 698-710.

Kellersberger, K. A.; Yu, E.; Kruppa, G. H.; Young, M. M.; Fabris, D. Top-down Characterization of Nucleic Acids Modified by Structural Probes Using High-Resolution Tandem Mass Spectrometry and Automated Data Interpretation. *Anal. Chem.* **2004**, *76*, 2438-2445.

Koc, H.; Tretyakova, N. Y.; Walker, V. E.; Henderson, R. F.; Swenberg, J. A. Molecular Dosimetry of N-7 Guanine Adduct Formation in Mice and Rats Exposed to 1,3-Butadiene. *Chem. Res. Toxicol.* **1999,** *12*, 566-574.

Koehl, P.; Burnouf, D.; Fuchs, R. P. P. Construction of Plasmids Containing a Unique Acetylaminofluorene Adduct Located within a Mutation Hot Spot: A New Probe for Frameshift Mutagenesis. *J. Mol. Bio.* **1989a**, *207*, 355-364.

Koehl, P.; Valladier, P.; Lefevere, J.-F.; Fuchs, R. P. P. Strong Structure Effect of the Position of a Single Acetyleaminofluorene Adduct Within a Mutation Hot Spot. *Nucleic Acids Res.* **1989b,** *17*, 9531-9541.

Kohn, K. W.; Hartley, J. A.; Mattes, W. B. Mechanism of DNA Sequence Selective Alkylation of Guanine-N7 Positions by Nitrogen Mustard. *Nucleic Acids Res.* **1987,** *15*, 10531-10549.

Koomen, J. M.; Russell, W. K.; Tichy, S. E.; Russell, D. H. Accurate Mass Measurement of DNA Oligonucleotide Ions Using High-Resolution Time-of-Flight Electrospray Mass Spectrometry. *J. Mass Spectrom.* **2002,** *37*, 357-371.

Kresbach, G. M.; Itani, M.; Saha, M.; Rogers, E. J.; Vouros, P.; Giese, R. W. Ester and Related Derivatives of Ring N-Pentafluorobenzylated 5-Hydroxymethyluracil: Hydrolytic Stability, Mass Spectral Properties, and Trace Detection by GC-ECD, GC-ECNI-MS and Moving Belt LC-ECNI-MS. *J. Chromatogr., B: Anal. Technol. Biomed. Life Sci.* **1989,** *476*, 424.

Kriek, E.; Miller, J. A.; Juhl, U.; Miller, E. C. 8-(N-2-Fluorenylacetamido) Guanosine, an Arylamidation Reaction Product of Guanosine and the Carcinogen N-Acetoxy-N-2-Fluorenylacetamide in Neutral Solution. *Biochemistry* **1967,** *1967*, 177-182.

Kuklenyik, Z.; Ekong, J.; Cutchins, C. D.; Needham, L. L.; Calafat, A. M. Simultaneous Measurement of Urinary Bisphenol A and Alkylphenols by Automated Solid-Phase Extractive Derivatization Gas Chromatography/Mass Spectrometry. *Anal. Chem.* **2003,** *75*, 6820-6825.

Kuo, H.-W.; Ding, W.-H. Trace Determination of Bisphenol A and Phytoestrogens in Infant Formula Powders by Gas Chromatography-Mass Spectrometry. *J. Chromatogr., A* **2004,** *1027*, 67-74.

Lao, Y.; Villalta, P. W.; Wang, M.; Cheng, G.; Sturla, S. J.; Hecht, S. S. Development of a Sensitive Method for Quantitation of DNA Pyridyloxobutylation Adducts Derived from the Tobacco Specific Nitrosamine, 4-(Methylnitrosamino)-1-(3-Pyridyl)-1-Butanone (NNK). *Abstracts of Papers,* 230th National Meeting of the American Chemical Society, Washington, DC, Aug 28-Sept 1, 2005; American Chemical Society: Washington, DC, 2005; TOXI-039.

Lao, Y.; Villalta, P. W.; Sturla, S. J.; Wang, M.; Hecht, S. S. Quantitation of Pyridyloxobutyl DNA Adducts of Tobacco-Specific Nitrosamines in Rat Tissue DNA by High-Performance Liquid Chromatography-Electrospray Ionization-Tandem Mass Spectrometry. *Chem. Res. Toxicol.* **2006a,** *19*, 674-682.

Lao, Y.; Yu, N.; Kassie, F.; Hecht, S. S. Formation and Accumulation of Pyridyloxobutyl (POB)-DNA Adducts in F344 Rats Treated with Tobacco-Specific Carcinogens. *Abstracts of Papers,* 232nd National Meeting of the American Chemical Society, San Francisco, CA, Sept 10-14, 2006; American Chemical Society: Washington, DC, 2006b; TOXI-031.

Lee, S. H.; Blair, I. A. Analysis of Etheno 2'-Deoxyguanosine Adducts as Dosimeters of Aldo-keto Reductase Mediated-Oxidative Stress. *ACS Symp. Ser.* **2004,** *865*, 139-152.

Lexa, M.; Valle, G. PRIMEX: Rapid Identification of Oligonucleotide Matches in Whole Genomes. *Bioinformatics* **2003**, *19*, 2486-2488.

Li, C.-S.; Wu, K.-Y.; Chang-Chien, G.-P.; Chou, C.-C. Analysis of Oxidative DNA Damage 8-Hydroxy-2'-Deoxyguanosine as a Biomarker of Exposures to Persistent Pollutants for Marine Mammals. *Environ. Sci. Technol.* **2005,** *39*, 2455-2460.

Li, K.-M.; Todorovic, R.; Devanesan, P.; Higginbotham, S.; Kofeler, H.; Ramanathan, R.; Gross, M. L.; Rogan, E. G.; Cavalieri, E. L. Metabolism and DNA inding Studies of 4-Hydroxyestradiol and Estradiol-3,4-Quinone in Vitro and in Female ACI Rat Mammary Gland in Vivo. *Carcinogenesis* **2004,** *25*, 289-297.

Liao, Q.; Chiu, N. H. L.; Shen, C.; Chen, Y.; Vouros, P. Investigation of Enzymatic Behavior of Benzonase/Alkaline Phosphatase in the Digestion of Oligonucleotides and DNA by ESI-LC/MS. *Anal. Chem.* **2007,** *79*, 1907-1917.

Liao, Q.; Shen, C.; Vouros, P. GenoMass - A Computer Software for Automated Identification of Oligonucleotide DNA Adducts from LC-MS Analysis of DNA Digests. *J. Mass Spectrom.*, 2009, *44*, 549-560.

Lynch, A. M.; Murray, S.; Boobis, A. R.; Davies, D. S.; Gooderham, N. J. The Measurement of MeIQx Adducts with Mouse Haemoglobin in Vitro and in Vivo: Implications for Human Dosimetry. *Carcinogenesis* **1991**, *12*, 1067-1072.

Manabe, A.; Kaneko, S.; Numazawa, S.; Itoh, K.; Inoue, M.; Hisamitsu, H.; Sasa, R.; Yoshida, T. Detection of Bisphenol-A in Dental Materials by Gas Chromatography-Mass Spectrometry. *Dent. Mater. J.* **2000,** *19*, 75-86.

Maragou, N. C.; Lampi, E. N.; Thomaidis, N. S.; Koupparis, M. A. Determination of Bisphenol A in Milk by Solid Phase Extraction and Liquid Chromatography-Mass Spectrometry. *J. Chromatogr., A* **2006,** *1129*, 165-173.

Markham, D. A.; McNett, D. A.; Birk, J. H.; Klecka, G. M.; Bartels, M. J.; Staples, C. A. Quantitative Determination of Bisphenol-A in River Water by Cool On-Column Injection-Gas Chromatography-Mass Spectrometry. *Int. J. Environ. Anal. Chem.* **1998,** *69*, 83-98.

Marzilli, L. A.; Wang, D.; Kobertz, W. R.; Essigmann, J. M.; Vouros, P. Mass Spectral Identification and Positional Mapping of Aflatoxin B1-Guanine Adducts in Oligonucleotides. *J. Am. Soc. Mass Spectrom.* **1998,** *9*, 676-682.

Marzilli, L. A.; Koertje, C.; Vouros, P. Capillary Electrophoresis-Mass Spectrometric Analysis of DNA Adducts. *Methods Mol. Biol. (Totowa, NJ, U.S.)* **2001,** *162*, 395-406.

Matter, B.; Malejka-Giganti, D.; Csallany, A. S.; Tretyakova, N. Quantitative Analysis of the Oxidative DNA Lesion, 2,2-Diamino-4-(2-Deoxy-B-D-Erythro-Pentofuranosyl)Amino]-5(2H)-Oxazolone (Oxazolone), in Vitro and in Vivo by Isotope Dilution-Capillary HPLC-ESI-MS/MS. *Nucleic Acids Res.* **2006,** *34*, 5449-5460.

Mattes, W. B.; Hartley, J. A.; Kohn, K. W. DNA Sequence Selectivity of Guanine-N7 Alkylation by Nitrogen Mustards. *Nucleic Acids Res.* **1986,** *14*, 2971-2987.

McDonald, E. S.; Randon, K. R.; Knight, A.; Windebank, A. J. Cisplatin Preferentially Binds to DNA in Dorsal Root Ganglion Neurons in Vitro and in Vivo: a Potential Mechanism for Neurotoxicity. *Neurobiol. Dis.* **2005,** *18*, 305-313.

McLuckey, S.; Habibi-Goudarzi, S. Decompositions of Multiply Charged Oligonucleotide Anions. *J. Am. Chem. Soc.* **1993,** *115*, 12085-12095.

McLuckey, S.; Habibi-Goudarzi, S. Ion Trap Tandem-Mass Spectrometry Applied to Small Multiply-Charged Oligonucleotides with a Modified Base. *J. Am. Soc. Mass Spectrom.* **1994,** *5*, 740-747.

McLuckey, S.; Berkel, G.; Glish, G. Tandem Mass Spectrometry of Small, Multiply Charged Oligonucleotides. *J. Am. Soc. Mass Spectrom.* **1992,** *3*, 60-70.

Meiss, G.; Friedhoff, P.; Hahn, M.; Gimadutdinow, O.; Pingoud, A. Sequence Preferences in Cleavage of dsDNA and ssDNA by the Extracellular Serratia Marcescens Endonuclease. *Biochemistry* **1995,** *34*, 11979-88.

Moller, L.; Hofer, T. 32P]ATP Mediates Formation of 8-Hydroxy-2'-Deoxyguanosine from 2'-Deoxyguanosine, a Possible Problem in the 32P-Postlabeling Assay. *Carcinogenesis* **1997,** *18*, 2415-2419.

Nestle, M.; Roberts, W. K. An Extracellular Nuclease from Serratatia Marcescens. *J. Biol. Chem.* **1969**, *244*, 5213-5218.

Ni, J.; Pomerantz, S.; Rozenski, J.; Zhang, Y.; McCloskey, J. Interpretation of Oligonucleotide Mass Spectra for Determination of Sequence Using Electrospray Ionization and Tandem Mass Spectrometry. *Anal. Chem.* **1996,** *68*, 1988-1999.

Oberacher, H.; Huber, C. G. Capillary Monoliths for the Analysis of Nucleic Acids by Ion-Pair Reverse-Phase Chromatography-Electrospray Ionization Mass Spectrometry. *Trends Anal. Chem.* **2002,** *21*, 166-173.

Oberacher, H.; Oefner, P. J.; Parson, W.; Huber, C. G. On-Line Liquid Chromatography-Mass Spectrometry: A Useful Tool for the Detection of DNA Sequence Variation. *Angew. Chem., Int. Ed.* **2001a,** *40*, 3828-3830.

Oberacher, H.; Parson, W.; Muhlmann, R.; Huber, C. G. Analysis of Polymerase Chain Reaction Products by On-line Liquid Chromatography-Mass Spectrometry for 5-1Genotyping of Polymorphic Short Tandem Repeat Loci. *Anal. Chem.* **2001b,** *73*, 5109-5115.

Oberacher, H.; Wellenzohn, B.; Huber, C. Comparative Sequencing of Nucleic Acid by Liquid Chromatography-Mass Spectrometry. *Anal. Chem.* **2002,** *74*, 211-218.

Otteneder, M.; Plastaras, J. P.; Marnett, L. J. Reaction of Malondialdehyde-DNA Adducts with Hydrazines-Development of a Facile Assay for Quantification of Malondialdehyde Equivalents in DNA. *Chem. Res. Toxicol.* **2002,** *15*, 312-318.

Phillips, D. H.; Farmer, P. B.; Beland, F. A.; Nath, R. G.; Poirier, M. C.; Reddy, M. V.; Turteltaub, K. W. Methods of DNA Adduct Determination and Their Application to Testing Compounds for Genotoxicity. *Environ. Mol. Mutagen.* **2000,** *35*, 222-233.

Poirier, M. C.; Beland, F. A. DNA Adduct Measurements and Tumor Incidence during Chronic Carcinogen Exposure in Animal Models: Implications for DNA Adduct-Based Human Cancer Risk Assessment. *Chem. Res. Toxicol.* **1992,** *5*, 749-55.

Pomerantz, S. C.; Kowalak, J. A.; McCloskey, J. A. Determination of Oligonucleotide Composition from Mass Spectrometrically Measured Molecular Weight. *J. Am. Chem. Soc.* **1993,** *4*, 204-209.

Premstaller, A.; Oberacher, H.; Huber, C. G. High-Performance Liquid Chromatography-Electrospray Ionization Mass Spectrometry of Single- and Double-Stranded Nucleic Acids Using Monolithic Capillary Columns *Anal. Chem.* **2000,** *72*, 4386-4393.

Qu, Y.; Scarsdale, N. J.; Tran, M.-C.; Farrell, N. P. Cooperative Effects in Long-Range 1,4 DNA-DNA Interstrand Cross-Links Formed by Polynuclear Platinum Complexes: an Unexpected Syn Orientation of Adenine Bases Outside the Binding Sites. *J. Biol. Inorg. Chem.* **2003,** *8*, 19-28.

Qu, Y.; Scarsdale, N. J.; Tran, M.-C.; Farrell, N. Comparison of Structural Effects in 1,4 DNA-DNA Interstrand Cross-links Formed by Dinuclear and Trinuclear Platinum Complexes. *J. Inorg. Biochem.* **2004,** *98*, 1585-1590.

Rajesh, M.; Ramesh, A.; Ravi, P. E.; Balakrishnamurthy, P.; Coral, K.; Punitham, R.; Sulochana, K. N.; Biswas, J.; Ramakrishnan, S. Accumulation of 8-Hydroxydeoxyguanosine and Its Relationship with Antioxidant Parameters in Patients with Eales' Disease: Implications for Antioxidant Therapy. *Curr. Eye Res.* **2003,** *27*, 103-110.

Rajesh, M.; Wang, G.; Jones, R.; Tretyakova, N. Stable Isotope Labeling-Mass Spectrometry Analysis of Methyl- and Pyridyloxobutyl-Guanine Adducts of 4-(Methylnitrosamino)-1-(3-Pyridyl)-1-Butanone in p53-Derived DNA Sequences. *Biochemistry* **2005,** *44*, 2197-2207.

Randerath, K.; Reddy, M. V.; Gupta, R. C. ^{32}P-labeling Test for DNA Damage. *Proc. Natl. Acad. Sci. U.S.A.* **1981,** *78*, 6126-6129.

Ricicki, E. M.; Luo, W.; Fan, W.; Zhao, L. P.; Zarbl, H.; Vouros, P. Quantification of N-(Deoxyguanosin-8-yl)-4-Aminobiphenyl Adducts in Human Lymphoblastoid TK6 Cells Dosed with N-Hydroxy-4-Acetylaminobiphenyl and Their Relationship to Mutation, Toxicity, and Gene Expression Profiling. *Anal. Chem.* **2006,** *78*, 6422-6432.

Rill, R. L.; Marsch, G. A. Sequence Preferences of Colalent DNA Binding by Anti-(+)- and Anti-(-)-Benzo[a]pyrene Diol Epoxide. *Biochemistry* **1990,** *25*, 6051-6058.

Rogozin, I. B.; Pavlov, Y. I. Theoretical Analysis of Mutation Hotspots and Their DNA Sequence Context Specificity. *Mutat. Res. Rev. Mutat. Res.* **2003,** *544*, 65-85.

Rogozin, I. B.; Malyarchuk, B. A.; Pavlov, Y. I.; Milanesi, L. From Context-Dependence of Mutations to Molecular Mechanisms of Mutagenesis. *Abstracts of Papers,* Pacific Symposium on Biocomputing, Kohala Coast, HI, Jan 4-5, 2005, 409-420.

Rozenski, J.; McCloskey, J. A. Determination of Nearest Neighbors in Nucleic Acids by Mass Spectrometry. *Anal. Chem.* **1999,** *71*, 1454-1459.

Rozenski, J.; McCloskey, J. A. SOS: A Simple Interactive Program for Ab Initio Oligonucleotide Sequencing by Mass Spectrometry. *J. Am. Soc. Mass Spectrom.* **2002,** *13*, 200-203.

Rozenski, J.; Crain, P.; McCloskey, J. A. The RNA Modification Database. *Nucleic Acids Res.* **1999**, *27*, 196-197.

Rusconi, F. GNU Polyxmass: A Software Framework for Mass Spectrometric Simulations of Linear (Bio)polymeric Analytes. *BMC Bioinfo.* **2006**, *7*, 226-242.

Sabbioni, G.; Neumann, H. G. Biomonitoring of Arylamines: Hemoglobin Adducts of Urea and Carbamate Pesticides. *Carcinogenesis* **1990**, *11*, 111-115.

Sambe, H.; Hoshina, K.; Hosoya, K.; Haginaka, J. Direct Injection Analysis of Bisphenol A in Serum by Combination of Isotope Imprinting with Liquid Chromatography-Mass Spectrometry. *Analyst (Cambridge, U.K.)* **2005,** *130*, 38-40.

Santella, R. M.; Gammon, M.; Terry, M.; Senie, R.; Shen, J.; Kennedy, D.; Agrawal, M.; Faraglia, B.; Zhang, F. DNA Adducts, DNA Repair Genotype/Phenotype and Cancer Risk. *Mutat. Res.* **2005,** *592*, 29-35.

Schrader, W.; Linscheid, M. Determination of Styrene Oxide Adducts in DNA and DNA Components. *J. Chromatogr., A* **1995,** *717*, 117-125.

Schrader, W.; Linscheid, M. Styrene Oxide DNA Adducts: in Vitro Reaction and Sensitive Detection of Modified Oligonucleotides Using Capillary Zone Electrophoresis Interfaced to Electrospray Mass Spectrometry. *Arch. Toxicol.* **1997,** *71*, 588-595.

Schulz, W.; Moore, K. Momentum Builds Against Bisphenol A. *Chem. Eng. News* **2008**, *86* (17): 11.

Seeberg, E.; Fuchs, R. P. P. Acetylaminofluorene Bound to Different Guanines of the Sequence -GGCGCC- is Excised with Different Efficiencies by the UvrABC Excision Nuclease in a Pattern Not Correlated to the Potency of Mutation Induction. *Proc. Natl Acad. Sci. U.S.A.* **1990**, *87*, 191-194.

Shao, H.; Li, J.; Shi, Y. A Study on the Adduct Characteristics of Styrene and DNA. *Zhonghua Laodong Weisheng Zhiyebing Zazhi* **2002,** *20*, 347-349.

Shao, B.; Han, H.; Hu, J.; Zhao, J.; Wu, G.; Xue, Y.; Ma, Y.; Zhang, S. Determination of Alkylphenol and Bisphenol A in Beverages Using Liquid Chromatography/Electrospray Ionization Tandem Mass Spectrometry. *Anal. Chim. Acta* **2005,** *530*, 245-252.

Shiota, G.; Maeta, Y.; Mukoyama, T.; Yanagidani, A.; Udagawa, A.; Oyama, K.; Yashima, K.; Kishimoto, Y.; Nakai, Y.; Miura, T.; Ito, H.; Murawaki, Y.; Kawasaki, H. Effects of Sho-Saiko-to on Hepatocarcinogenesis And 8-Hydroxy-2'-Deoxyguanosine Formation. *Hepatology (Philadelphia, PA, U.S.)* **2002,** *35*, 1125-1133.

Singh, R.; McEwan, M.; Lamb, J. H.; Santella, R. M.; Farmer, P. B. An Improved Liquid Chromatography/Tandem Mass Spectrometry Method for the Determination of 8-Oxo-7,8-Dihydro-2'-Deoxyguanosine in DNA Samples Using Immunoaffinity Column Purification. *Rapid Commun. Mass Spectrom.* **2003,** *17*, 126-134.

Singh, R.; Gaskell, M.; Le Pla, R. C.; Kaur, B.; Azim-Araghi, A.; Roach, J.; Koukouves, G.; Souliotis, V. L.; Kyrtopoulos, S. A.; Farmer, P. B. Detection and Quantitation of Benzo[a]pyrene-Derived DNA Adducts in Mouse Liver by Liquid Chromatography-Tandem Mass Spectrometry: Comparison with 32P-Postlabeling. *Chem. Res. Toxicol.* **2006,** *19*, 868-878.

Soglia, J. R.; Turesky, R. J.; Paehler, A.; Vouros, P. Quantification of the Heterocyclic Aromatic Amine DNA Adduct N-(Deoxyguanosin-8-yl)-2-Amino-3-Methylimidazo[4,5-f]Quinoline in Livers of Rats Using Capillary Liquid Chromatography/Microelectrospray Mass Spectrometry: A Dose-Response Study. *Anal. Chem.* **2001,** *73*, 2819-2827.

Song, R.; Zhang, W.; Chen, H.; Ma, H.; Dong, Y.; Sheng, G.; Zhou, Z.; Fu, J. Site Determination of Phenyl Glycidyl Ether-DNA Adducts Using High-Performance Liquid

Chromatography with Electrospray Ionization Tandem Mass Spectrometry. *Rapid Commun. Mass Spectrom.* **2005,** *19*, 1120-1124.

Stiborova, M. R., M.; Hodek, P.; Frei, E.; Schmeiser, H. H. Monitoring of DNA Adducts in Humans and ^{32}P-Postlabelling Methods: A Review. *Collect. Czech. Chem. Commun.* **2004,** *69*, 476-498.

Straub, K. M.; Burlingame, A. L. Mass Spectrometry as a Tool for the Analysis of Xenobiotic-Modified Nucleic Acids. *Soft Ionization Biological Mass Spectrometry: Proceedings of the Chemical Society Symposium on Advances in Mass Spectrometry Soft Ionization Methods, London, July 1980*; Morris. H. R., Ed.; Heyden: London, 1981; pp 39-52.

Sturla, S. J.; Scott, J.; Lao, Y.; Hecht, S. S.; Villalta, P. W. Mass Spectrometric Analysis of Relative Levels of Pyridyloxobutylation Adducts Formed in the Reaction of DNA with a Chemically Activated Form of the Tobacco-Specific Carcinogen 4-(Methylnitrosamino)-1-(3-Pyridyl)-1-Butanone. *Chem. Res. Toxicol.* **2005,** *18*, 1048-1055.

Svec, F.; Huber, C. G. Monolithic Materials: Promise, Challenge, Achievements. *Anal. Chem.* **2006,** *78*, 2100-2107.

Tarun, M.; Rusling, J. F. Measuring DNA Nucleobase Adducts Using Neutral Hydrolysis and Liquid Chromatography-Mass Spectrometry. *Crit. Rev. Eukaryotic Gene Expression* **2005a,** *15*, 295-316.

Tarun, M.; Rusling, J. F. Quantitative Measurement of DNA Adducts Using Neutral Hydrolysis and LC-MS. Validation of Genotoxicity Sensors. *Anal. Chem.* **2005b,** *77*, 2056-2062.

Tarun, M.; Bajrami, B.; Rusling, J. F. Genotoxicity Screening Using Biocatalyst/DNA Films and Capillary LC-MS/MS. *Anal. Chem.* **2006,** *78*, 624-627.

Tebbs, R. S.; Romano, L. J. Mutagensis at a Site-Specifically Modified NarI Sequence by Acetylated and Deacetylated Aminofluorene Adducts. *Biochemistry* **1994,** *33*, 8998-9006.

Tominaga, T.; Negishi, T.; Hirooka, H.; Miyachi, A.; Inoue, A.; Hayasaka, I.; Yoshikawa, Y. Toxicokinetics of Bisphenol A in Rats, Monkeys and Chimpanzees by the LC-MS/MS Method. *Toxicology* **2006,** *226*, 208-217.

Topinka, J.; Schwarz, L. R.; Kiefer, F.; Wiebel, F. J.; Gajdos, O.; Vidova, P.; Dobias, L.; Fried, M.; Sram, R. J.; Wolff, T. DNA Adduct Formation in Mammalian Cell Cultures by Polycyclic Aromatic Hydrocarbons (PAH) and Nitro-PAH in Coke Oven Emission Extract. *Mutat. Res.* **1998,** *419*, 91-105.

Tost, J.; Gut, I. G. DNA Analysis by Mass Spectrometry - Past, Present and Future. *J. Mass. Spectrom.* **2006,** *41*, 981-995.

Tretyakova, N. Stable Isotope Labeling-Mass Spectrometry to Map the Reactivity of DNA Bases towards Carcinogens and Antitumor Agents. *Abstracts of Papers,* 226th National Meeting of the American Chemical Society, New York, NY, Sept 7-11, 2003; American Chemical Society: Washington, DC, 2003; TOXI-003.

Tretyakova, N. Y.; Ling, Y.-P.; Upton, P. B.; Sangaiah, R.; Swenberg, J. A. Macromolecular Adducts of Butadiene. *Toxicology* **1996,** *113*, 70-76.

Tretyakova, N.; Matter, B.; Jones, R.; Shallop, A. Formation of Benzo[a]pyrene Diol Epoxide-DNA Adducts at Specific Guanines within K-Ras and p53 Gene Sequences: Stable Isotope-Labeling Mass Spectrometry Approach. *Biochemistry* **2002,** *41*, 9535-9544.

Tsukioka, T.; Brock, J.; Graiser, S.; Nguyen, J.; Nakazawa, H.; Makino, T. Determination of Trace Amounts of Bisphenol A in Urine by Negative-Ion Chemical-Ionization-Gas Chromatography/Mass Spectrometry. *Anal. Sci.* **2003,** *19*, 151-153.

Tu, C.; Wu, X.; Liu, Q.; Wang, X.; Xu, Q.; Guo, Z. Crystal Structure, DNA-Binding Ability and Cytotoxic Activity of Platinum(II) 2,2'-Dipyridylamine Complexes. *Inorg. Chim. Acta* **2004,** *357*, 95-102.

Turesky, R. J.; Vouros, P. Formation and Analysis of Heterocyclic Aromatic Amine-DNA Adducts in Vitro and in Vivo. *J. Chromatogr., B: Anal. Technol.Biomed. Life Sci.* **2004,** *802*, 155-166.

Turesky, R. J.; Freeman, J. P.; Holland, R. D.; Nestorick, D. M.; Miller, D. W.; Ratnasinghe, D. L.; Kadlubar, F. F. Identification of Aminobiphenyl Derivatives in Commercial Hair Dyes. *Chem. Res. Toxicol.* **2003,** *16*, 1162-1173.

Uematsu, Y.; Hirata, K.; Suzuki, K.; Iida, K.; Saito, K. Chlorohydrins of Bisphenol A Diglycidyl Ether (BADGE) and of Bisphenol F Diglycidyl Ether (BFDGE) in Canned Foods and Ready-to-Drink Coffees from the Japanese Market. *Food Addit. Contam.* **2001,** *18*, 177-185.

Upadhyaya, P.; Sturla, S. J.; Tretyakova, N.; Ziegel, R.; Villalta, P. W.; Wang, M.; Hecht, S. S. Identification of Adducts Produced by the Reaction of 4-(Acetoxymethylnitrosamino)-1-(3-Pyridyl)-1-Butanol with Deoxyguanosine and DNA. *Chem. Res. Toxicol.* **2003,** *16*, 180-190.

Van den Driessche, B.; Lemiere, F.; Van Dongen, W.; Esmans, E. L. Alkylation of DNA by Melphalan: Investigation of Capillary Liquid Chromatography-Electrospray Ionization Tandem Mass Spectrometry in the Study of the Adducts at the Nucleoside Level. *J. Chromatogr., B: Anal. Technol. Biomed. Life Sci.* **2003,** *785*, 21-37.

Van den Driessche, B.; Lemiere, F.; Van Dongen, W.; Esmans, E. L. Structural Characterization of Melphalan Modified 2'-Oligodeoxynucleotides by Miniaturized LC-ES MS/MS. *J. Am. Soc. Mass Spectrom.* **2004a,** *15*, 568-579.

Van den Driessche, B.; Lemiere, F.; Van Dongen, W.; Van der Linden, A.; Esmans, E. L. Qualitative Study of in Vivo Melphalan Adduct Formation in the Rat by Miniaturized Column-Switching Liquid Chromatography Coupled with Electrospray Mass Spectrometry. *J. Mass Spectrom.* **2004b,** *39*, 29-37.

Vanhoutte, K.; Van Dongen, W.; Hoes, I.; Lemiere, F.; Esmans, E. L.; Van Onckelen, H.; Van den Eeckhout, E.; van Soest, R. E. J.; Hudson, A. J. Development of a Nanoscale Liquid Chromatography/Electrospray Mass Spectrometry Methodology for the Detection and Identification of DNA Adducts. *Anal. Chem.* **1997,** *69*, 3161-3168.

Villalta, P. W.; Hecht, S. S.; Sturla, S. J.; Wang, M.; Scott, J.; Lao, Y.; Cheng, G. Investigating DNA Adduct Formation from Tobacco Smoke Exposure using Mass Spectrometry. *Abstracts of Papers,* 61st Northwest Regional Meeting of the American Chemical Society, Reno, NV, June 25-28, 2006; American Chemical Society: Washington, DC, 2006; RE06-221.

Vrkic, A. K.; O'Hair, R. A. J.; Foote, S.; Reid, G. E. Fragmentation Reactions of All 64 Protonated Trimer Oligonucleotides and 16 Mixed Base Tetramer Oligonucleotides via Tandem Mass Spectrometry in an Ion Trap. *Int. J. Mass Spectrom.* **2000**, *19,* 145-164.

Vulimiri, S. V.; DiGiovanni, J. DNA Adducts as Biomarkers of DNA Damage in Lung Cancer. In *Biomarkers of Environmentally Associated Disease: Technologies, Concepts, and Perspectives;* Wilson, S. H., Suk, W. A., Eds.; Lewis Publishers: Boca Raton, FL, 2002; pp. 109-126.

Wang, J.; Xiong, S.; Xie, C.; Markesbery, W. R.; Lovell, M. A. Increased Oxidative Damage in Nuclear and Mitochondrial DNA in Alzheimer's Disease. *J. Neurochem.* **2005,** *93*, 953-962.

Wang, M.; McIntee, E. J.; Cheng, G.; Shi, Y.; Villalta, P. W.; Hecht, S. S. A Schiff Base Is a Major DNA Adduct of Crotonaldehyde. *Chem. Res. Toxicol.* **2001,** *14*, 423-430.

Wang, M.; Cheng, G.; Sturla, S. J.; Shi, Y.; McIntee, E. J.; Villalta, P. W.; Upadhyaya, P.; Hecht, S. S. Identification of Adducts Formed by Pyridyloxobutylation of Deoxyguanosine and DNA by 4-(Acetoxymethylnitrosamino)-1-(3-Pyridyl)-1-Butanone, a Chemically Activated Form of Tobacco Specific Carcinogens. *Chem. Res. Toxicol.* **2003,** *16*, 616-626.

Wang, Y.; Taylor, J.-S.; Gross, M. L. Nuclease P1 Digestion Combined with Tandem Mass Spectrometry for the Structure Determination of DNA Photoproducts. *Chem. Res. Toxicol.* **1999,** *12*, 1077-1082.

Warnke, U.; Rappel, C.; Meier, H.; Kloft, C.; Galanski, M.; Hartinger, C. G.; Keppler, B. K.; Jaehde, U. Analysis of Platinum Adducts with DNA Nucleotides and Nucleosides by Capillary Electrophoresis Coupled to ESI-MS: Indications of Guanosine 5'-Monophosphate O6-N7 Chelation. *ChemBioChem* **2004,** *5*, 1543-1549.

Willems, A. V.; Deforce, D. L.; Van den Eeckhout, E. G.; Lambert, W. E.; Van Peteghem, C. H.; De Leenheer, A. P.; Van Bocxlaer, J. F. Analysis of Benzo[a]pyrene Diol Epoxide-DNA Adducts by Capillary Zone Electrophoresis-Electrospray Ionization-Mass Spectrometry in Conjunction with Sample Stacking. *Electrophoresis* **2002,** *23*, 4092-4103.

Willems, A. V.; Deforce, D. L.; Van Peteghem, C. H.; Van Bocxlaer, J. F. Analysis of Nucleic Acid Constituents by On-line Capillary Electrophoresis-Mass Spectrometry. *Electrophoresis* **2005,** *26*, 1221-1253.

Wolf, S. M.; Vouros, P. Application of Capillary Liquid Chromatography Coupled with Tandem Mass Spectrometric Methods to the Rapid Screening of Adducts Formed by the Reaction of N-Acetoxy-N-Acetyl-2-Aminofulorene with Calf Thymus DNA. . *Chem. Res. Toxicol.* **1994,** *7*, 82-88.

Wolf, S. M.; Vouros, P. Incorporation of Sample Stacking Techniques into the Capillary electrophoresis CF-FAB Mass Spectrometric Analysis of DNA Adducts. *Anal. Chem.* **1995,** *67*, 891-900.

Wu, K.-Y.; Scheller, N.; Ranasinghe, A.; Yen, T.-Y.; Sangaiah, R.; Giese, R.; Swenberg, J. A. A Gas Chromatography/Electron Capture/Negative Chemical Ionization High-Resolution Mass Spectrometry Method for Analysis of Endogenous and Exogenous N7-(2-Hydroxyethyl) Guanine in Rodents and Its Potential for Human Biological Monitoring. *Chem. Res. Toxicol.* **1999,** *12*, 722-729.

Xi, Z.; Chao, F.; Sun, Y.; Li, G.; Liu, Z.; Li, Y.; Yang, D.; Zhang, H.; Zhang, W. Analysis on Diesel Vehicle Exhaust and Its Oxidative DNA Damage. *Zhongguo Gonggong Weisheng* **2004,** *20*, 419-421.

Xiong, W.; Glick, J.; Lin, Y.; Vouros, P. Separation and Characterization of Diastereoisomeric Oligonucleotide Adducts Using Monolithic Columns by Ion-Pair Reversed-Phase Nano-HPLC Coupled to Ion Trap Mass Spectrometry. *Anal. Chem.* **2007,** *79*, 5312-5321.

Xu, J.; Blair, I. A. Quantitative Analysis of DNA Adducts Generated through Nonenzymatic Pathways Using LC/MS. *Abstracts of Papers,* 228th National Meeting of the American Chemical Society, Philadelphia, PA, Aug 22-26, 2004; American Chemical Society: Washington, DC, 2004; TOXI-093.

Yang, J.; Wang, B.; Rusling, J. F. Genotoxicity Sensor Response Correlated with DNA Nucleobase Damage Rates Measured by LC-MS. *Mol. Biosystems* **2005,** *1*, 251-259.

Yue, G.; Luo, Q.; Zhang, J.; Wu, S.-L.; Karger, B. L. Ultratrace LC/MS Proteomic Analysis Using 10-mm-i.d. Porous Layer Open Tubular Poly (Styrene-Divinylbenzene) Capillary Columns. *Anal. Chem.* **2007,** *79*, 938-946.

Ziegel, R.; Shallop, A.; Jones, R.; Tretyakova, N. K-Ras Gene Sequence Effects on the Formation of 4-(Methylnitrosamino)-1-(3-Pyridyl)-1-Butanone (NNK)-DNA Adducts. *Chem. Res. Toxicol.* **2003,** *16*, 541-550.

Ziegel, R.; Shallop, A.; Upadhyaya, P.; Jones, R.; Tretyakova, N. Endogenous 5-Methylcytosine Protects Neighboring Guanines from N7 and O6-Methylation and O6-Pyridyloxobutylation by the Tobacco Carcinogen 4-(Methylnitrosamino)-1-(3-Pyridyl)-1-Butanone. *Biochemistry* **2004,** *43*, 540-549.

VDM publishing house ltd.

Scientific Publishing House

offers

free of charge publication

of current academic research papers, Bachelor´s Theses, Master's Theses, Dissertations or Scientific Monographs

If you have written a thesis which satisfies high content as well as formal demands, and you are interested in a remunerated publication of your work, please send an e-mail with some initial information about yourself and your work to *info@vdm-publishing-house.com.*

Our editorial office will get in touch with you shortly.

VDM Publishing House Ltd.
Meldrum Court 17.
Beau Bassin
Mauritius
www.vdm-publishing-house.com

VDM Verlag Dr. Müller
LAP LAMBERT Academic Publishing
SVH Südwestdeutscher Verlag für Hochschulschriften

Made in the USA
Middletown, DE
09 April 2024

52769003R00091